SpringerBriefs in Applied Sciences and Technology

More information about this series at http://www.springer.com/series/8884

Adi Wolfson · Shlomo Mark
Patrick M. Martin · Dorith Tavor

Sustainability through Service

Perspectives, Concepts and Examples

 Springer

Adi Wolfson
Patrick M. Martin
Dorith Tavor
Green Processes Center
Sami Shamoon College of Engineering
Beer-Sheva
Israel

Shlomo Mark
Negev Monte Carlo Research Center
(NMCRC)
Sami Shamoon College of Engineering
Beer-Sheva
Israel

ISSN 2191-530X ISSN 2191-5318 (electronic)
ISBN 978-3-319-12963-1 ISBN 978-3-319-12964-8 (eBook)
DOI 10.1007/978-3-319-12964-8

Library of Congress Control Number: 2014954595

Springer Cham Heidelberg New York Dordrecht London

Springer is part of Springer Science+Business Media (www.springer.com)

Preface

Scientific research and practice in the complementary fields of service and sustainability sciences have garnered increased attention in recent years, and as such, both have been extensively developed. The many parallels between service and sustainability—both require broad and comprehensive vision and perspectives and the integration of knowledge, methods, tools and skills from different disciplines, from the exact and engineering sciences to social science and the humanities—have demonstrated the synergistic potential of their co-development. Yet, although the service sector has grown exponentially to become the most dominant sector of the market and aspects of sustainability have become integral factors in decision-making processes, the exploration of their mutual relationship, i.e., sustainable services and sustainability as a service, is still in its infancy.

This book outlines the main paradigms, concepts and terminology in the fields of sustainability and service science, in the process discussing the benefits to be gained by—and the challenges entailed in—learning to maximize the potential of their reciprocal relationship. It begins with a brief review in each field separately, which is followed by a literature survey focused on the integration of sustainability and service and that continues by proposing some primary methodologies and measures to assess and engineer sustainable services. Finally, it offers a novel outlook, according to which sustainability is conceptualized as a service.

We wish to acknowledge the support of Sami Shamoon College of Engineering, whose support enabled us to perform in-depth research in the field of sustainable service, which was necessary to complete the book.

Adi Wolfson
Shlomo Mark
Patrick M. Martin
Dorith Tavor

Contents

Chapter 1
Sustainability

Abstract The well-known and growing environmental problems of climate change and biodiversity loss on a global scale have compelled humanity to reconsider its mutual relationship with the natural environment. Indeed, since at least the late 20th century, the search for solutions to the challenges associated with human population expansion within the restricted space and using the limited amount of resources afforded by Earth has driven people toward more environmentally conscious development, i.e., sustainable development. Sustainability integrates the environmental, social and economic dimensions of life over large scales of time and space (i.e., from the short- to long-term and from the local to the global scale, respectively). As such, sustainable practices are those that incorporate environment, society and economy in the production and delivery of goods and services while simultaneously generating added value, making a profit, and meeting the needs of both current and future generations. To translate theory into action and effect a real change, however, the decision-making process must be imbued with sustainability. Therefore, the adoption and implementation of sustainable practices relies on the design of simple, reliable and comparative measures for their evaluation and the development of the appropriate frameworks and methods to realize them in practice.

1.1 Background

Have you ever thought about the origins of the water you use in the morning to brush your teeth and wash your face, or where the wastewater you produce ends up? And what about the amount of water that was used during this process, or the amounts of water and energy needed to prepare your cup of coffee? Also, were the groceries for your breakfast grown locally or imported from abroad, and which chemicals were used during their growth? Similar questions can be asked about every good or service that is used on a regular, daily basis, because regardless of

© The Author(s) 2015

A. Wolfson et al., *Sustainability through Service*, SpringerBriefs in Applied Sciences and Technology, DOI 10.1007/978-3-319-12964-8_1

whether it is a necessity or a luxury, its production and delivery consumes natural resources and is associated with emissions, effluent, and waste.

In the not so distant past, when people or families were self-sufficient and produced and supplied most of their daily needs on their own, they knew how to evaluate the exact weight and value of any product, including the materials, energy, and effort consumed during the production process, and they appreciated the abundant resources provided by the natural world. Moreover, the hierarchy between human being and nature was clear, and there was a general belief that the world was governed by a certain order, managed and controlled, as it were, by a *force majeure*. The Industrial Revolution of the 18th and 19th centuries, however, marked a shift in manufacturing and agriculture from long and tedious hand labor using manually operated tools and machines to mass production based on power-driven machines and assembly lines [1]. It catapulted societal development and urbanization and, in terms of goods and services, changed how people produced and traded goods, transforming forever people's daily lives. Yet it also fundamentally altered how people related to nature, weakening the belief in a *force majeure*, and in the process it threatened to change the natural world order. Once the industrial revolution was firmly established, various religious and political movements combined with the rapid technological progress of the day [2] to further accelerate urbanization [3], industrialization [4] and cultural consumption [5], a process that only widened the gap even more between people and nature. Moreover, nowadays most of our consumption is supplied through outsourcing, and therefore, its value is translated into a price tag that has to be paid without knowing the real value of each product. Indeed, the progress that humanity has made has come at a heavy cost, entailed in the marked degradation of our social and natural environments, and it has raised many questions about the relationship between human beings and nature and about the ability of nature and its resources to continue sustaining life as we know it.

Simultaneously inspirational and challenging, the natural world has for millennia evoked the curiosity of scientists and curious laypeople alike, driving them to investigate and mimic a wide variety of biological, chemical and physical phenomena and processes. However, though its fundamentals were already laid down by Hippocrates and Aristotle, *ecology*—the scientific study of interactions among organisms and their environment that incorporates elements from the fields of biology and earth science—only became a separate science in the late 19th century [6, 7]. With respect to people's actions and their influence on the natural environment, ecology proposed a general model of *conservation* [8], which historically has been rearward-oriented and focused on the *preservation* of natural areas and of the *biodiversity* contained therein [9]. As such, ecology paralleled the social approach that was taken at this time with regard to preserving the enchantment of certain, usually older, cultures and even the buildings associated with those cultures. Yet this perspective differentiated between production processes and natural activities, and even allocated different areas for each activity. In so doing, it failed to recognize that the environment, the society and the economy had to be understood holistically, as an integrated whole.

Later on, mainly in the 20th century, the combination of air pollution from the widespread burning of coal to support the heavily industrialized world and the other hazardous chemicals being discharged at that time into the environment became a deadly threat to the health of many groups in society. The potential risks inherent in such wanton pollution constituted the main incentive behind the establishment of environmental societies and the involvement of environmentalists in politics. Thus, during the 1960s and 1970s, grass roots environmental movements emerged across the globe, often forming important parts of local and national politics, and eventually, environmental issues were included as integral parts of decision-making processes. At the same time and with the understanding that people's actions affect both the social and the natural environments, for good and for bad, an integrated, quantitative, and interdisciplinary approach to the study of the interaction between the social and natural environments and to the search for solutions to environmental problems emerged in the form of *environmental science*. Yet this model was mainly focused on the *"end of the pipe"* dangers and hazards created in production processes and generated by daily living and emitted to the environment in myriad forms of pollution [10, 11].

However, at the end of the 20th and the beginning of 21st centuries, as environmentalism continued to grow in popularity and gain recognition on a global scale, the focus moved from preservation and "end of the pipe" treatments to the *"rational use of resources"* and *"thinking and planning ahead"*, i.e., the intentional design of more *environmentally friendly* or *green* production processes [12]. Valuable insight has been obtained into how to design and develop goods and services differently, in the process stimulating the development of new strategies and fields like *green chemistry* [13], *green engineering* [14], *green building* [15] and even *green fashion* [16]. Concomitantly, the demand for more environmentally conscious, responsible and effective production processes has grown, accelerating the quest for and implementation of *clean technologies* (*Cleantechs*) that can help realize this goal [16, 17].

Finally, these new perspectives also penetrated the economy, which today is the main driving force behind most decisions made and actions taken. The debate as to whether nature is part of the economy or vice versa has evolved to appreciate that the two can develop together. As such, the belief that economic growth—which facilitates opportunities for market expansion—can be integrated with social justice and welfare and with protection of the environment, i.e., *green growth*, has taken hold [18, 19].

The paradigm of *sustainability* that emerged in the 1970s grew out of the integration of the environment, society and the economy in a manner that allows for profit. The novel approach was intended to serve as a basis for a framework that defined how to design, develop, implement and manage processes while creating and maintaining the conditions under which humans can coexist with the natural world in productive harmony [20–24]. However, the failure to fully implement those ideas of 30 years ago has led to the catastrophic environmental problems that today threaten our existence, such as climate change and water and air pollution, which, in turn, have rejuvenated efforts in the 21st century to realize sustainability.

Insofar as it is a measure of the capacity of a system to balance the needs of the present without compromising the ability of future generations to meet their own needs, sustainability bridges between body and soul and between humanity and nature. Necessarily multidisciplinary, it combines intelligence, feeling, knowledge,

Fig. 1.1 The dimensions of
sustainability

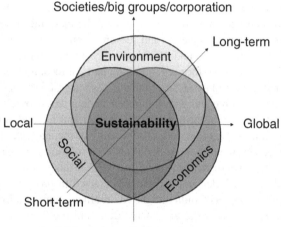

love and care in an approach based on comprehensive, long-range planning that, as previously mentioned, accounts for the continuity and integral interrelationship of the economic, social, and environmental aspects of life [57]. In addition, sustainability must consider the local and global as well as the short- and long-term effects on individuals and societies (Fig. 1.1). Viewing sustainability from this perspective stresses the importance of incorporating doctrines that promote and propagate sustainability on a daily basis and in all areas of life. To do so entails a continuous process of learning and creatively adapting to the change inherent in life. Accordingly, the processes involved in maintaining sustainability are complex and not straightforward. Indeed, one of the main obstacles to applying sustainability concepts in practice involves effecting a shift from theory to practice, part of which entails breaking the theoretical concepts down into more practical, visible, and accessible values, methods, and tools.

One of the main channels for achieving sustainability is via *sustainable development* [26, 27]. Introduced by the Brundtland Commission of the United Nations (UN) World Commission on Environment and Development in its report in 1987 entitled Our Common Future, sustainable development is defined as "development that meets the needs of the present without compromising the ability of future generations to meet their own needs" [28]. It combines environmental, economic, and social concerns, incorporating them into values, skills, and operational tools, which necessarily involves elements from the natural, engineering and technology sciences as well as from management and the behavioral sciences. This approach is meant to insure that the final result is a stable, realistic, widely applicable and fair process.

A central milestone in the development of the notion of sustainability that promoted general awareness and that elicited proposals for action was the United Nations Conference on Environment and Development (UNCED—*Earth Summit)* held in Rio de Janeiro, Brazil, in 1992. Out of that conference emerged an action agenda for the 21st century, i.e., *Agenda 21* [25]. A framework for action to realize sustainable development at the local, national, and global levels, Agenda 21 was adopted by the UN and by governments and organizations around the world. In addition, the UN established the Commission on Sustainable Development to implement some of the recommendations in Agenda 21. Furthermore, since 1992, the UN and other organizations have held several conferences and published numerous declarations reinforcing their commitments to Agenda 21 and defining how it should be implemented. Since 1992, however, Agenda 21 has elicited growing opposition that is focused mainly in the United States, where people are concerned that the implementation of sustainability measures may deprive individuals of their property rights.

As the world's population is constantly growing in parallel with a general increase in quality of life (and the associated consumption of natural resources and damage to the environment), the ideas set forth in Agenda 21 have inspired new sustainability models that transcend the traditional definitions of sustainability. Their overarching goal is to devise a system that will enable people to live rich lives that are not restricted by rigid demands to survive on only the bare essentials while maintaining the natural and social environments. New concepts that have emerged include (i) *liveability*, which refers to the suitability of a place to human habitation, (ii) *resilience*, or the capability to prepare for and respond to environmental threats [29], and (iii) *regeneration*, which has at its heart the concept of "*net zero*" or the goal to use only the amounts of resources that can be produced renewably, i.e., living in a closed economy. In addition, a vareity of new concepts and paradigms, such as *share economy* and *smart cities*, inherently included sustainability features in their design and development and in their implementation.

1.2 Sustainability Science

The broad objectives and multidisciplinary nature of the concept of sustainability entailed in the requirement that it integrate the data, knowledge, skills and abilities of experts from a variety of fields and disciplines, motivated the World Congress on Challenges of a Changing Earth 2001 in Amsterdam to introduce the research discipline of *sustainability science* [30]. Assigning a clear definition to this new paradigm, however, has proven to be a challenge. Over the last decade, therefore, sustainability science has been repeatedly redefined. Common to the definitions, however, is an emphasis on the need to consolidate features from different knowledge systems into practical methods and tools that can be practically applied to promote sustainability on a worldwide scale.

Kieffer et al. [31] suggested that sustainability science is "the cultivation, integration, and application of knowledge about earth systems gained especially from the holistic and historical sciences (such as geology, ecology, climatology and oceanography) coordinated with knowledge about human interrelationships gained from the social sciences and humanities, in order to evaluate, mitigate, and minimize the consequences, regionally and worldwide, of human impacts on planetary systems and on societies across the globe and into the future—that is, in order that humans can be knowledgeable earth stewards". Clark and Dickson [32] added that ideally, it "brings together scholarship and practice, global and local perspectives from north and south, and disciplines across the natural and social sciences, engineering, and medicine", and Reitan [33] argued that it "must encompass different magnitudes of scales (of time, space, and function), multiple balances (dynamics), multiple actors (interests) and multiple failures (systemic faults)".

1.3 Sustainability Measures

The old business adage "if you can't measure it, you can't manage it" is highly applicable today to sustainability. Moreover, as the UN Commission on Sustainable Development argued, "we measure what we value, and value what we measure" [34]. Thus, to implement sustainability practice and to define whether an action, a process, a good or a service is sustainable, there is a need to quantitatively and/or qualitatively measure sustainability using simple, reliable and comparative indicators, indexes and measures.

In general, indicators can translate the knowledge of the physical and social sciences into manageable units of information to guide and facilitate decision-making processes and to help communicate ideas and measure changes and progress. The history of using indicators as tools to measure the overall wellbeing of a given nation began in the late 1940s when the concepts of gross domestic product (GDP) and gross national product (GNP) were introduced. Although over the years there has been considerable objection to their use, GDP and GNP are still used today to assess progress. In 1987, the Brundtland Commission stressed the need to devise new measures of progress that will go beyond those based exclusively on economic issues to capture a more holistic sense of human and ecological wellbeing, i.e., the essence of sustainable development [35]. In Chap. 40 of Agenda 21, therefore, the commission articulated a call on countries to develop and identify—at the national, international, governmental and NGO levels—indicators of sustainable development that can provide a solid basis for decision-making at all levels [34]. Such efforts to define sustainability indicators have attracted much criticism, however, vis-à-vis the challenge of encapsulating myriad complex and diverse processes into a few and relatively simple measures. As such, it is clear that to advance sustainability practices, the systems or processes involved must be simplified. Thus, while on the one hand sustainability indicators should capture the key issues and their complex interrelationships into understandable and manageable numbers, there is an acute awareness, on the other hand, that such endeavors entail trade-offs that could ultimately result in the creation of misleading indicators.

1.3.1 Measures

Different sustainability measures have been introduced over the years to measure the sustainability of products, processes, and countries, to name a few [36–38]. In general, sustainability indicators are divided between the three components of sustainability—environmental, social and economic—but because of the sheer number of parameters and indicators, it is important that we identify the most significant and representative primary indicators, i.e., core sustainability indicators. Table 1.1 lists some of the commonly used and accepted indicators and indexes for calculating the environmental, economic, and social impacts of a good or a service.

Table 1.1 Representative indicators and indexes

	Indicator/Index	Explanation
Environmental (EN)	Carbon footprint	Total amount of greenhouse gas emissions from a process
	Air quality	A measure of the condition of the air relative to the requirements of one or more biotic species
	Waste generation	Amount of waste that is generated per process
	Consumption of ozone depleting substances	Amounts of ozone-depleting substances being eliminated as per the Montreal Protocol
	Forest area as a percent of land area	Areas of natural and plantation forest tracked over time
	Arable and permanent crop land area	Total area of "arable land" and "land under permanent crops"
	Use of agricultural pesticides	Use of pesticides per unit of agricultural land area
Economic (EC)	Process gross domestic product	Measure of the economic performance of a process
	Employment	Number of employees per process
	Income	Median income per capita
	Balance of trade in goods and services	Difference between the value of exported goods and services and the value of imported goods and services
	Intensity of material use	Quantities of minerals and metals, including primary and secondary (recycled) materials, consumed per unit of real GDP
	Annual energy consumption per capita	Per capita amount of energy— liquids, solids, gases and electricity—available in a given year in a given country or geographical area
	Intensity of energy use in transportation	Energy consumption for transportation relative to the amount of freight or passengers carried and the distance traveled
	Rate of waste recycling and reuse	Volume of waste reused or recycled based on the volume actually generated at source on a per capita basis

(continued)

Table 1.1 (continued)

	Indicator/Index	Explanation
Social (SO)	Human development index	Measure of the level of human development in the country
	Health impact	Potential effects of a process on the health of a population and the distribution of those effects within the population
	Unemployment rate	Ratio of number of unemployed people to the size of the labor force
	Nutritional status of children	Children under age five whose weight-for-age and height-for-age values are either between 80 and 120 % of the reference value of the country or within two standard deviations of this value
	Number of recorded Crimes per 100,000 population	Total crimes recorded in criminal (police) statistics, regardless of type
	Population growth rate	Average annual rate of change of population size during a specified period

One of the most useful and well-known indicators of sustainability is the *carbon footprint*, which is a measure of the total greenhouse gas emissions from a process or during the production of a product [39–41]. Since every process involves the use of materials and energy that can be equivalently expressed via carbon dioxide emissions, the carbon footprint represents the total environmental impact of a process or a product and enables comparisons on a single scale between different processes/products. This important measure has gained wide acceptance, and in 2006 the Carbon Trust Company of the UK introduced a new service, named *carbon labeling*, to add a product's carbon footprint to its label just as, for example, calorie count or price is clearly stated [42, 43]. Yet carbon labeling also sparked a debate about the meaning of labeling, the boundaries of the system and the reliability of the numbers [44]. Furthermore, although carbon labeling, for example, allows one to choose between alternatives, it does not reflect all the factors that contribute to sustainability, such as the social conditions or salaries of the employees who make the products. Moreover, one cannot determine, based on the overall carbon label alone, in which part of the production process of a given good or service one should invest to effect the largest reduction in the carbon footprint.

As a single indicator is usually unable to provide a sufficiently representative picture to facilitate valid decision-making about sustainability, several indexes and ratings, each of which comprises a number of indicators that are standardized into a single scale, have been proposed in recent years to measure sustainability [45]. For example, the *air quality health index*, an indication on a scale of 1–10 of the level of health risk associated with local air quality, is calculated based on several air pollutant indicators such as carbon monoxide and particulate emissions as well as ozone level [46]. Another example is the *ecological footprint*, which quantifies the equivalent of land required to maintain the lifestyle of a population or of a person or which provides a measure of the intrinsic sustainability of urban areas. Included in the ecological footprint are indicators that measure the consumption of natural resources, the discharge of effluents and the refuse treatment systems in use in a particular region [47].

As a standardized measure, the ecological footprint can be used to comparatively analyze different countries. Nevertheless, the methods of calculation and the ultimate effectiveness of the ecological footprint have come under heavy fire, mostly due to the plethora of socio-economic factors that vary across countries but also to the fact that the measure considers neither the quantity nor the quality of consumption. Alternatively, the *environmental sustainability* and *environmental performance indexes* were suggested to quantify the environmental sustainability and performance of countries. These measures are based not only on natural resource use, but also on the effectiveness of this use and on policies regarding environmental issues [48, 49]. Common to all such measures, however, is that they render the environmental aspect of sustainability redundant and do not allow for the identification and examination of which factors are the most important determinants of sustainability or, correspondingly, of where investments should be made to effect real change.

Finally, sustainability can be translated into more widely accepted economic measures to enable its easy quantification by assigning monetary values to sustainability assessments. In so doing, sustainability can be clearly expressed as a universally accessible parameter that can be used to reach policy making decisions. One such new measure, *Energypoints* [50], equates resource consumption with gallons of gasoline, a value that can be easily converted to any currency. Another example is *sustainable value*, a measure used by companies to assess their sustainability performance, a calculation of how much more or less return a company generated with its particular set of economic, environmental and social resources compared to a benchmark, i.e., it reflects the company's efficiency of resource use [51, 52].

1.3.2 Sustainability Assessment

To confront the complexities associated with measuring sustainability, a variety of methodologies have been used. In general, each can be assigned to one of three different measurement doctrines [53–56]: (i) *Reductionist* or *top-down* strategies break complex processes down into their components and study how these work in isolation and then together; (ii) *Participatory* or *bottom-up* strategies are based on the social sciences and set goals and priorities within local contexts while actively promoting research to stimulate action or change; (iii) *Adaptive learning processes* or *integrated methodologies* are hybridizations between bottom-up and top-down strategies to obtain levels of knowledge sufficient to provide a deeper understanding of the interaction between environmental, social and economic systems.

Fiksel et al. [57] suggested four fundamental principles that sustainability assessments should follow: (i) Address the dual perspectives of resource consumption and value creation, (ii) include economic, environmental, and social aspects, (iii) systematically consider each stage in the product life-cycle, and (iv) develop both leading and lagging indicators. In addition, sustainability measures must integrate the impacts of the environmental, economic, and social dimensions with considerations of the local and global as well as the short- and long-term effects. For example, while an investment in solar cells to produce solar energy has a clear environmental benefit, it may also have straightforward social benefits if the solar energy is generated for areas that are not connected to an electrical grid. Furthermore, over time the initial investment in the solar technology can be recovered until eventually the solar power plant generates an economic profit, thus justifying that initial investment, as the profit can subsequently be used to support the plant's continued operation. Moreover, an investment in solar energy to promote sustainability should also consider its impact with respect to the location of the power plant, since its potential impact in sunny countries in Africa, for example, will be higher than in northern latitudes where sun exposure is much lower. Likewise, the production of a certain cloth in China or in the United States must account not only for local variations in fabric prices, but also for the different environmental and social regulations in place in the two countries. As such, the production processes and labor laws of each country translate into differences in the working environments and salaries of the employees. But current measures for assessing the sustainability of a good or service, like the carbon footprint, consider only the process, based on energy and materials use, and do not distinguish between different uses of the product or the salaries, levels of education and number of employees that were involved in its production. For example, if based solely on carbon footprint, the purchase of a can of Coca-Cola ($\sim$170 g CO_2 for 330 ml aluminum can or 0.52 g CO_2/g) is better than buying a loaf of bread ($\sim$1,000 g CO_2 per loaf or 1.25 g CO_2/g), even though the value of bread in terms of our food requirements is much higher. In addition, regarding the fabric production example, if the processes used in the two countries to manufacture the cloth are the same, the carbon footprint does not differentiate between them, despite the gross differences in prices, local labor regulations, and levels of social equity between the two countries.

One of the most accepted and widely used sustainability assessment methods is the *life-cycle assessment* or *analysis* (LCA). Although it originated in the field of biology, the concept of life-cycle—or a cyclic process—is not limited in its applicability to natural processes. Indeed, it is also used widely in the economics and management fields and to describe tangible products, i.e., goods, for which it applies to everything involved in product realization, from its design and development process to its launch into the market until eventually the market stabilizes and the product is considered mature. In addition to the management or commercial life-cycle of a product, there is also its resource-oriented life-cycle, which traces the utilization of energy and materials—from the point they enter the production process to their use and disposal until they are finally recycled or recovered—involved in the manufacture of the product. The LCA approach, devised in the 1980s to quantify the environmental impact of a product during its entire life span, can be used to identify opportunities for improvement [58, 59]. However, despite its promising potential, the LCA still must overcome certain challenges, the first of which is to develop an accepted "standard" methodology for its measurement.

Several known and acceptable frameworks for the assessment and reporting of sustainability exist. One of the first, introduced in 1997 by John Elkington, modified the conventional business term "bottom line", which usually refers to the performance of a business in terms of its net profit or loss, to "triple bottom line", in which are incorporated measures for social equity and environmental concerns and which is based on full cost accounting, i.e., sustainable profit or loss [60, 61]. Another approach to imbuing the standard business strategy with sustainability is corporate sustainability [62–64], which can be described as "meeting the needs of a firm's direct and indirect stakeholders (such as shareholders, employees, clients, pressure groups, communities, etc.) without compromising its ability to meet the needs of future stakeholders as well" [62]. Recently, a new method, termed circles of sustainability, was suggested to assess the sustainability of cities and to investigate the problems typically faced by communities or organizations. Based on a combination of qualitative and quantitative indicators, circles of sustainability integrate the four domains of economics, ecology, politics, and culture [65].

References

1. Hobsbawm EJ (1999) Industry and empire: from 1750 to the present day. Penguin Books, London
2. Hanson FA (2013) Technology and cultural tectonics: shifting values and meanings. Palgrave Macmillan, New York
3. Timberlake M (1985) Urbanization in the world-economy. Academic Press, London
4. Kiely R (2005) Industrialization and development. Routledge, Abingdon
5. Storey J (1999) Cultural consumption and everyday life. Oxford University Press, Oxford
6. Kormondy EJ (1969) Concepts of ecology. Prentice-Hall Inc., New Jersey
7. Begon M, Townsend CR, Harper JL (2009) Ecology: from individuals to ecosystems. Wiley, New York
8. Allen R (1980) How to save the world. Strategy for world conservation. Kogan Page Ltd., London
9. Groombridge B (1992) Global biodiversity: status of the earth's living resources. Chapman & Hall, London
10. O'Riordan T (1995) Environmental science for environmental management. Longman Group Limited, New York
11. Kupchella CE, Hyland MC (1993) Environmental Science: living within the system of nature. Prentice Hall International, New Jersey
12. Leff E (1995) Green production: toward an environmental rationality. Guilford Press, New York
13. Anastas PT, Warner JC (2000) Green chemistry: theory and practice. Oxford University Press, Oxford
14. Allen DT, Shonnard DR (2001) Green engineering: environmentally conscious design of chemical processes. Pearson Education, New Jersey

15. Kibert CJ (2012) Sustainable construction: green building design and delivery. Wiley, New Jersey
16. Pernick R, Wilder C (2007) The clean tech revolution: the next big growth and investment opportunity. Harper-Collins Publishers, New York
17. Pernick R, Wilder C (2008) The clean tech revolution: discover the top trends, technologies, and companies to watch. Harper-Collins Publishers, New York
18. Fletcher K, Grose L, Hawken P (2012) Fashion and sustainability: design for change. Laurence King Ltd., London
19. Ekins P (2002) Economic growth and environmental sustainability: the prospects for green growth. Routledge, London
20. Hallegatte S (2012) From growth to green growth—a framework. No. w17841. National Bureau of Economic Research, Cambridge
21. Kates RW, Clark WC (1999) Our common journey: a transition toward sustainability. National Academies Press, Washington DC
22. Willard B (2002) The sustainability advantage. New Society Publishers, Gabriola Island
23. Edwards AR (2005) The sustainability revolution: portrait of a paradigm shift. New Society Publishers, Gabriola Island
24. Dresner S (2008) The principles of sustainability, 2nd edn. EarthScan, Oxford
25. Rogers PP, Jalal KF, Boyd JA (2008) An introduction to sustainable development. EarthScan, Oxford
26. Dalal-Clayton BDD, Bass S (2002) Sustainable development strategies: a resource book, vol 1. OECD Publishing, Paris
27. http://www.un.org/documents/ga/res/42/ares42-187.htm. Accessed 25 May 2014
28. Robinson NA (1993) Agenda 21: earth's action plan. Oceana Publications Inc., Oxford
29. Coaffee J (2008) Risk, resilience, and environmentally sustainable cities. Energy Policy 36(12):4633–4638
30. Kates RW, Clark WC, Corell R, Hall JM, Jaeger CC, Lowe I, McCarthy JJ, Schellnhuber HJ, Bolin B, Dickson NM, Faucheux S, Gallopin GC, Grübler A, Huntley B, Jäger J, Jodha NS, Kasperson RE, Mabogunje A, Matson P, Mooney H, Moore B, O'Riordan T, Svedin U (2001) Sustainability science. Science 292(5517):641–642
31. Kieffer SW, Barton P, Palmer AR, Reitan PH, Zen E (2003) Megascale events: natural disasters and human behavior. Geol Soc America abstracts with programs 432
32. Clark WC, Dickson NM (2003) Sustainability science: the emerging research program. Proc Nat Acad Sci USA 100(14):8059–8061
33. Reitan P (2005) Sustainability science and what's needed beyond science. Sustain Sci Pract Policy 1(1):77–80
34. United Nations Commission of Sustainable Development (2001) Indicators of sustainable development: guidelines and methodologies. United Nations Publications, New York
35. Hardi P, Zdan T (1997) Assessing sustainable development: principles in practice. International Institute for Sustainable Development, Canada
36. Rennings K, Wiggering H (1997) Steps towards indicators of sustainable development: linking economic and ecological concepts. Ecol Econ 20(1):25–36
37. Hák T, Moldan B, Dahl AL (2007) Sustainability indicators: a scientific assessment. Island Press, Washington DC
38. United Nations Department of Economic (2007) Indicators of sustainable development: guidelines and methodologies. United Nations Publications, New York
39. Wiedmann T, Minx J (2007) A definition of 'carbon footprint'. Ecol Econ Res Trends 2:55–65
40. Benjaafar S, Li Y, Daskin M (2013) Carbon footprint and the management of supply-chains: insights from simple models. IEEE Trans Autom Sci Eng 10(1):99–116
41. Pandey D, Agrawal M, Pandey JS (2011) Carbon footprint: current methods of estimation. Environ Monit Assess 178(1–4):135–160
42. http://www.carbontrust.com/. Accessed 25 May 2014

43. Edwards-Jones G, Plassmann K, York EH, Hounsome B, Jones DL, Milà CL (2009) Vulnerability of exporting nations to the development of a carbon label in the United Kingdom. Environ Sci Policy 12(4):479–490
44. Boardman B (2008) Carbon labelling: too complex or will it transform our buying? Significance 5(4):168–171
45. Rajesh KS, Murty HR, Gupta SK, Dikshit AK (2009) An overview of sustainability assessment methodologies. Ecol Indic 9(2):189-212
46. http://www.ec.gc.ca/cas-aqhi/default.asp?lang=En&n=79A8041B-1. Accessed 25 May 2014
47. Wackernagel M, Rees WE (2013) Our ecological footprint: reducing human impact on the earth. No. 9. New Society Publishers, Gabriola Island
48. Färe R, Grosskopf S, Hernandez-Sancho F (2004) Environmental performance: an index number approach. Resour Energy Econ 26(4):343–352
49. http://epi.yale.edu/. Accessed 25 May 2014
50. http://www.energypoints.com/company/. Accessed 25 May 2014
51. http://www.sustainablevalue.com/. Accessed 25 May 2014
52. Figge F, Hahn T (2008) Sustainable investment analysis with the sustainable value approach—a plea and a methodology to overcome the instrumental bias in socially responsible investment research. Prog Ind Ecol 5(3):255–272
53. Reed MS, Fraser EDG, Morse S, Dougill AJ (2005) Integrating methods for developing sustainability indicators to facilitate learning and action. Ecol Soc 10(1):r3
54. Pretty JN (1995) Participatory learning for sustainable agriculture. World Dev 23(8):1247–1263
55. Reed MS, Fraser EDG, Dougill AJ (2006) An adaptive learning process for developing and applying sustainability indicators with local communities. Ecol Econ 59(4):406–418
56. Bell S, Morse S (2008) Sustainability indicators: measuring the immeasurable? Earthscan, Oxford
57. Fiksel J, McDaniel J, Mendenhall C (1999) Measuring progress towards sustainability principles: process and best practices. Battelle Memorial Institute, Ohio
58. Kloepffer W (2008) Life cycle sustainability assessment of products. Int J Life Cycle Assess 13(2):89–95
59. Guinée JB, Heijungs R (2005) Life cycle assessment. Wiley, New Jersey
60. Slaper TF, Hall TJ (2011) The triple bottom line: what is it and how does it work? Indiana Business Review 86(1):4–8
61. Wayne N, MacDonald C (2004) Getting to the bottom of "triple bottom line". Bus Ethics Q 243–262
62. Thomas D, Hockerts K (2002) Beyond the business case for corporate sustainability. Bus Strategy Environ 11(2):130–141
63. Wilson M (2003) Corporate sustainability: what is it and where does it come from. Ivey Bus J 67(6):1–5
64. Salzmann O, Ionescu-Somers A, Steger U (2005) The business case for corporate sustainability: literature review and research options. Eur Manag J 23(1):27–36
65. http://citiespro.pmhclients.com/images/uploads/Indicators_-_Briefing_Paper.pdf. Accessed 25 May 2014

Chapter 2
Service

Abstract An intangible product, service is a value that is produced and delivered simultaneously from provider to customer. The delivery and operation of a service rely on a variety of assets, processes and activities that may include, alone or in combination, people, knowledge, resources, methods, processes, and technologies. In addition, the service value is normally co-created jointly and reciprocally in interactions between provider and customer, using a constellation of integrated resources and capabilities that both parties share, combine and renew. Recently, a new paradigm was proposed that aims to overcome many of the limitations inherent in "goods-dominant (G-D) logic" for thinking about commerce, marketing and exchange. Known as "service-dominant (S-D) logic", it suggests that the focus be moved from the exchange of goods to that of services, and that in fact, all exchanges between producers and consumers are based on services. Finally, as service research has become multidisciplinary, including concepts from marketing, computer science, information systems, and operations as well psychology and sociology, a new discipline entitled service science, whose goal is to create a platform for systematic service innovation, was advanced.

2.1 Definition

Although the term "service" is used extensively today, across the different disciplines, such as marketing, operations, and computer science, it has correspondingly different meanings and connotations [1]. The dictionary entry for "*service*" comprises two main descriptions: the action of helping or doing work for someone and the organized system of apparatuses, appliances, employees, etc., to supply a public need. The search for a more professional and precise definition yielded two other explanations: a type of economic activity that is intangible, is not stored and does not result in ownership and, simply, an *intangible product*.

In general, the concept of service can be described as the transformation of *value*, an intangible product, from the service *supplier* (also termed the *provider*)

© The Author(s) 2015

A. Wolfson et al., *Sustainability through Service*, SpringerBriefs in Applied
Sciences and Technology, DOI 10.1007/978-3-319-12964-8_2

Fig. 2.1 Schematic
representation of service

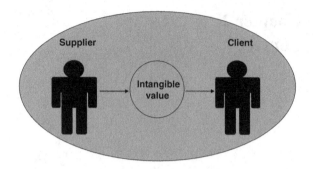

to the *client* (also termed the *consumer* or *customer*) (Fig. 2.1) [1, 2]. The process
of transformation can be set in motion by a client whose needs can be provided by
the supplier, by a service supplier who offers a particular service to the client, or
through a combination of the two actions. In contrast to the supply of a good, the
consumption of which occurs after production, the supply of a service involves the
simultaneous production and consumption of value [3]. Therefore, service is also
described in terms of value creation [4]. Usually not an individual action, however,
the creation of value entailed in each service delivery is driven by a configura-
tion of people, resources, information, and technology, to name but a few of the
important elements of a service. As such, terms like "*value-chain*" and "*service-
system*", which reflect service structure and organization, are often used in service
research, innovation, management and marketing [1, 5]. Likewise, because they
are not stand-alone entities, services usually achieve their goals more effectively
by interacting with other products, i.e., goods or services, and processes, i.e.,
manufacturing and agriculture. Lastly, delivered and consumed on a daily basis in
every sector of the economy, services can be classified by their motivation, theme,
type or initiator.

In general, services can be divided into three main groups: (i) *Intellectual and
spiritual*, such as the services provided by synagogues and churches, (ii) *behavioral*, for instance, using public transportation, and (iii) *practical or operational*,
as in consulting an insurance agency [6]. Table 2.1 lists the themes and types of
services.

2.2 Service Characteristics

Although the discussion in the early services marketing literature (up to 1980)
was mainly conceptual [7], a few empirical studies later examined how consumers differentiate between goods and services [8] and strived to develop services
marketing strategies [9]. Later still, services were proposed to have four main
characteristics—intangibility, inseparability, perishability and heterogeneity—to
illustrate the unique nature of services, distinguish between goods and services
and offer a basis for services marketing and innovation [10–12].

Table 2.1 Representative examples of services

Field	Type	Example
Consulting	Education	Education planning
	Health	Medical opinion
	Economic	Investment
Maintaining	Education	Continuing education program
	Health	Health insurance
	Economic	Loan
Administrative	Education	School registration
	Health	Surgery appointment
	Economic	Bank account open
Information	Education	Online lecture
	Health	Online blade test results
	Economic	Bank account printout

The *intangibility* of services means that in contrast to goods, services cannot be seen or touched or experienced before their delivery. It also dictates that they cannot be stored or owned by either the supplier or the client, although the value of a service can be transferred from a supplier to a client. For example, a flight service that is delivered by an airline to its passengers cannot belong to the passengers, and it is impossible for them to know ahead of time whether their use of the service will be a good or a bad experience.

Services are also characterized by *inseparability*, as they are simultaneously delivered and consumed, which also requires client participation in the process, thereby allowing or even obliging the client to have an effect on the process. For example, while concrete goods (e.g., furniture, automobile, etc.) can be purchased after their production and without meeting their producers, in contrast, the use of a service such as a personal consulting service requires that the consultant produce the information and knowledge at the same time that it is delivered and consumed by the client.

Related to the intangibility of services is their *perishability*—i.e., because services cannot be stored and all tangible and intangible resources and systems involved in their realization are assigned for a definite time during service delivery, services are therefore irreversible, which suggests they are also time dependent. For example, while groceries can be purchased at a certain time but used later or even returned to the store, a service such as that delivered by a diet consultant is produced and delivered simultaneously, and regardless of whether the information was used, it cannot be returned to the dietitian.

Finally, services are also characterized by *heterogeneity* or *inconsistency*, reflecting the potentially wide variability of any service. As such, the delivery of a service can never be repeated in exactly the same way, as the supplier, the customer and the place and time change from one delivery to the next. For example, the manufacture of a computer, whether done by people in Israel or in the US, yields exactly the same product. The quality and characteristics of the service later delivered by the computer salesperson, however, will depend both on the knowledge and skills of the service supplier and on the knowledge and needs of the customer, and as such, the potential for variability is high.

2.3 Service Structure

Taking the whole process point of view, the phases of a service can be illustrated based on a life-cycle structure, from its initiative and design stage through production and delivery to maintenance and retirement. Each of these phases, in turn, comprises several activities or processes [13]. Portrayed linearly, the *service life-cycle* at its most basic entails the phases of initiative (initiation and requirements), design (modeling, identify values and evaluation), construction (build and compose, integration and processing), testing (validation and evaluation), operation (deployment, provision and management), maintenance and retirement (Fig. 2.2).

The service life-cycle is a strategy that supports service organizations and helps them undergo healthy growth and development and to recognize their operational potential. As an indicator of overall service development, the life-cycle assessment can be used to analyze the service from a variety of different perspectives, such

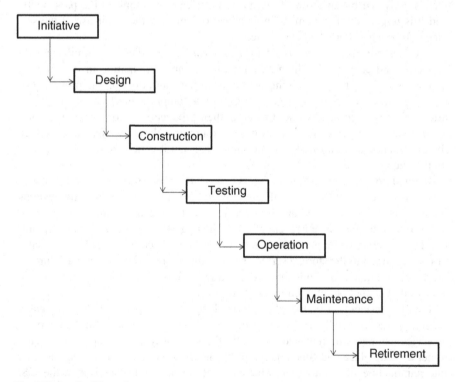

Fig. 2.2 Service life-cycle

as economics or cost minimization, optimal performance or quality, and based on behavioral elements as well as resource utilization (for more on the concept of life-cycle assessment, see Sect. 1.3.2). Thus, initiative and design, the preliminary phase of the service life-cycle, has as its goal the planning and optimization of the service strategy to support the service's goals and objectives. The service production and delivery phase guides the service to facilitate its efficient and effective transition from its planning and strategy phase to its operation phase, while simultaneously integrating with other services or processes. Finally, the service maintenance and retirement phase describes the practical aspects of day-to-day operations, maintenance and support that keep things running smoothly, reliably, efficiently and cost-effectively.

2.4 Service Design Methodologies

Although services have already been produced and used for many years, the systematic and organized design of services has been incorporated into a relatively new discipline [14]. While service design must comprise creativity, innovation and artistic ability, on the one hand, on the other, it should be practicable and applicable, and it should also consider the economic, social and psychological ramifications of the service. In general, however, like any other design scheme, service design is a set of guidelines, principles and techniques that can be effectively (re)organized and (re)deployed to support and enable strategic plans and productivity [15–17]. In addition, the design process should consider the entire life-cycle of the service. As such, it should usually begin by identifying a need or a problem and then continue by generating a solution, i.e., consolidating knowledge, capabilities and resources to deliver high-quality, workable, service-oriented solutions. In addition, the maintenance of the service, its integration with other services and products, and its evolution and re-design should also be considered from the life-cycle perspective.

2.5 Service Delivery and Use

The process of service delivery, and especially the operation of a service, relies on a variety of assets, processes and activities that may include, alone or in combination, people, knowledge, resources, methods, processes, and technologies. For example, consulting with a doctor can be delivered either in the *person-to-person mode*, through an appointment at the doctor's office, or via a dedicated Internet site. In addition, each service is a composite of several important factors and of defined, measurable, and practicable values and methods and the alignment of intangible and tangible products. Thus, each value can be described as a chain of entities that connects the supplies and the end-user or customer, i.e., *supply-chain*.

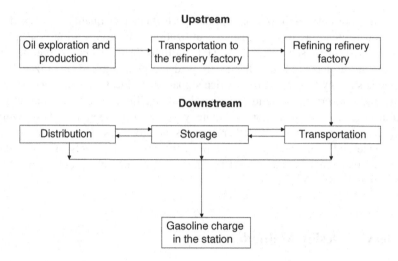

Fig. 2.3 Fuel supply-chain

Figure 2.3 illustrates the example of an automobile fuel supply-chain, from oil exploration and production through transportation to the refinery, where it is refined for use in transportation, and on to its distribution and storage until it is delivered to the gasoline station and supplied to the customer. Of course, this is a generalized scheme, and one should bear in mind that it can be extensively elaborated by expanding each of the building blocks into their detailed entities.

Finally, in many cases, a certain process, such as service delivery or the production of a good, can be divided into several actions/entities, each of which can be described as a basic service. Known as *service-oriented architecture*, this approach has gained increased attention in recent years in terms of the innovation, research and management of businesses. Today, the service-oriented architecture is implemented mainly in the design of computer software [18, 19].

2.6 Value Creation

The nature of value and its creation has been discussed and debated for many years. As the essence of each service entails the delivery of value from the supplier to the client, the various types of values must be defined. In general, two approaches to the creation and delivery of value, *value in-exchange* (Fig. 2.4a) and *value in-use* (Fig. 2.4b), are recognized. In a value-in-exchange model, the value is created by the supplier and delivered simultaneously to the client who consumes that value. In this case, the client is actually a *consumer* who has no impact on the value creation process, and in that respect, this type of model simulates the goods production and delivery model. Thus, although for a service the production and delivery of value occur concomitantly, they are, in fact, two separate processes. An

Fig. 2.4 Value delivery
models: **a** value in-exchange
and **b** value in-use

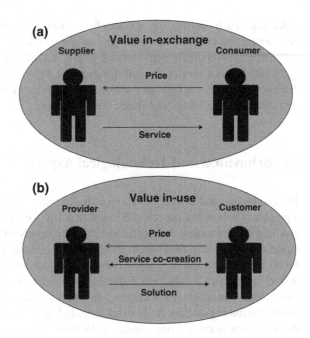

example of this type of service is the purchase of prescription medicine in a pharmacy, where the pharmacist supplies the client with the doctor-prescribed medication and the client pays for it. Another example is the public transportation sector, in which the supplier of a bus service creates the platform, comprising buses, drivers, stations, routes and timetables, and the client utilizes and pays for the service.

In contrast, a value-in-use model conceptualizes services as shared and dynamic problem-solving endeavors that create value in multiple dimensions. As such, the *supplier* becomes a *provider* that produces and delivers the service to the customer by providing the platform, which entails the resources, facilities, effort and knowledge required for value delivery. But this perspective also encourages the greater involvement of the client, who shares in the production and delivery of the service, in the process transforming from mere consumer to *customer* [20, 21]. Services of this type include consultation with a pharmacist about the different brands of a medication and about how to use it and the use of a taxi service, which in contrast to public transportation, the exact time and place that a taxi is ordered are determined by the customer.

Thus, value in-use actually means that the value is always *co-created*, jointly and reciprocally, in interactions among providers and beneficiaries using a constellation of integrated resources and capabilities that provider and customer share, combine and renew [21]. In general, service *co-creation* can be divided into three types [22]: (i) *consumption* or *co-usage*—the customers exploit a service and passively co-create value by creating the perception of value, for example, the use of a full-service gasoline station, where the provider supplies and delivers the service, (ii) *co-performance* or *co-production*—the customers share in some of the tasks required to deliver the service, for instance, the self-use of a gasoline station, where the customers use the gas station to refill their tanks themselves, and (iii) *co-design*—a dialog between customers and service providers determines the types and form of service desired, for example, the design of a gas station service that provides both the full- and self-service refilling modes in the same place, but also other services like convenience store services.

2.7 Behavioral and Technological Aspects of Service

The production and delivery of services is also affected by cultural, social, psychological and behavioral aspects [23] as well as by the change inherent in technologies and their ongoing development [24]. The development of new technologies, especially those that are computer- and Internet-based, have forever changed the nature of services, and they allow both the provider and the customer to create and deliver the service's value via different routes and in more efficient, flexible and cost-effective manners. Furthermore, the ubiquity and sophistication of new information technologies like the Internet have led to fundamental, ongoing change in the ways that organizations interact with their customers [25]. Finally, new technologies and changes in the way that we create and consume products, i.e., goods and services, have revealed that the opportunities to design and develop new services are virtually unlimited [26].

The combination of new technologies, particularly information technologies and the Internet, together with the innovation of the co-creation principle, have also resulted in the development of new service modes such as: (i) *super-service*— mainly performed by the supplier, (ii) *self-service*—the customer takes an active part and invests knowledge, skills and facilities to execute most of the service, and (iii) *mixed-service*—both provider and customer share most of the tasks and capabilities [27–30] (Fig. 2.5). Using flight reservations as an example, in a super-service mode, reserving flight tickets, which includes the search for a flight on the desired date and the final ordering of the places on the flight, is done solely by the travel agency based on the customer's request. Customers who prefer the self-service mode use the Internet platform of the agency themselves to search for and purchase their tickets. And customers who exploit a mixed-service mode preform a preliminary search of the Internet themselves for the flights they want and then reserve the tickets through the services of a travel agency.

2.8 Service Card

As noted in the sections above, a variety of components should be considered and identified in the design and development of a service. Figure 2.6 summarizes these elements in a *service card*.

Resources/Tasks/Capabilities

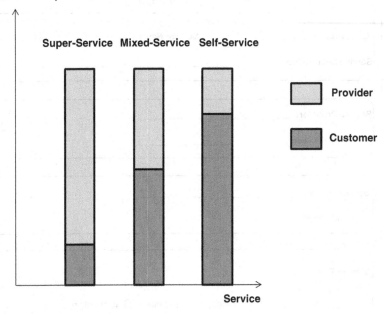

Fig. 2.5 Various modes of services

2.9 Service Performance: Measures and Indicators

Service performance evaluations rely on appropriate methods for the measurement and analysis of performance and on effective approaches that can be exploited to optimize performance. In general, service performance is evaluated in terms of productivity, efficiency and quality, which reflect on service organization, customer satisfaction, the mutual trust that develops between provider and customer, and the profits generated by the service [31–35]. In addition, as the assessment of service performance and the development of general indicators while simultaneously integrating all the relevant factors constitute a highly complex process, service performance methodologies and measurements bear the strongest resemblance to tangible products.

The productivity of a service is defined as a ratio of the output of a service unit to its input, while efficiency measures the ability to transform input to output. As such, service productivity evaluations entail parameters such as time of production and delivery and time of use as well as the technologies and resources used in the realization of the service. On the other hand, service quality, as a measure of how well a service conforms to the client's expectations, can be used to improve the service and increase client satisfaction. Thus, while both service productivity and efficiency are measures of service performance from the point of view and for the

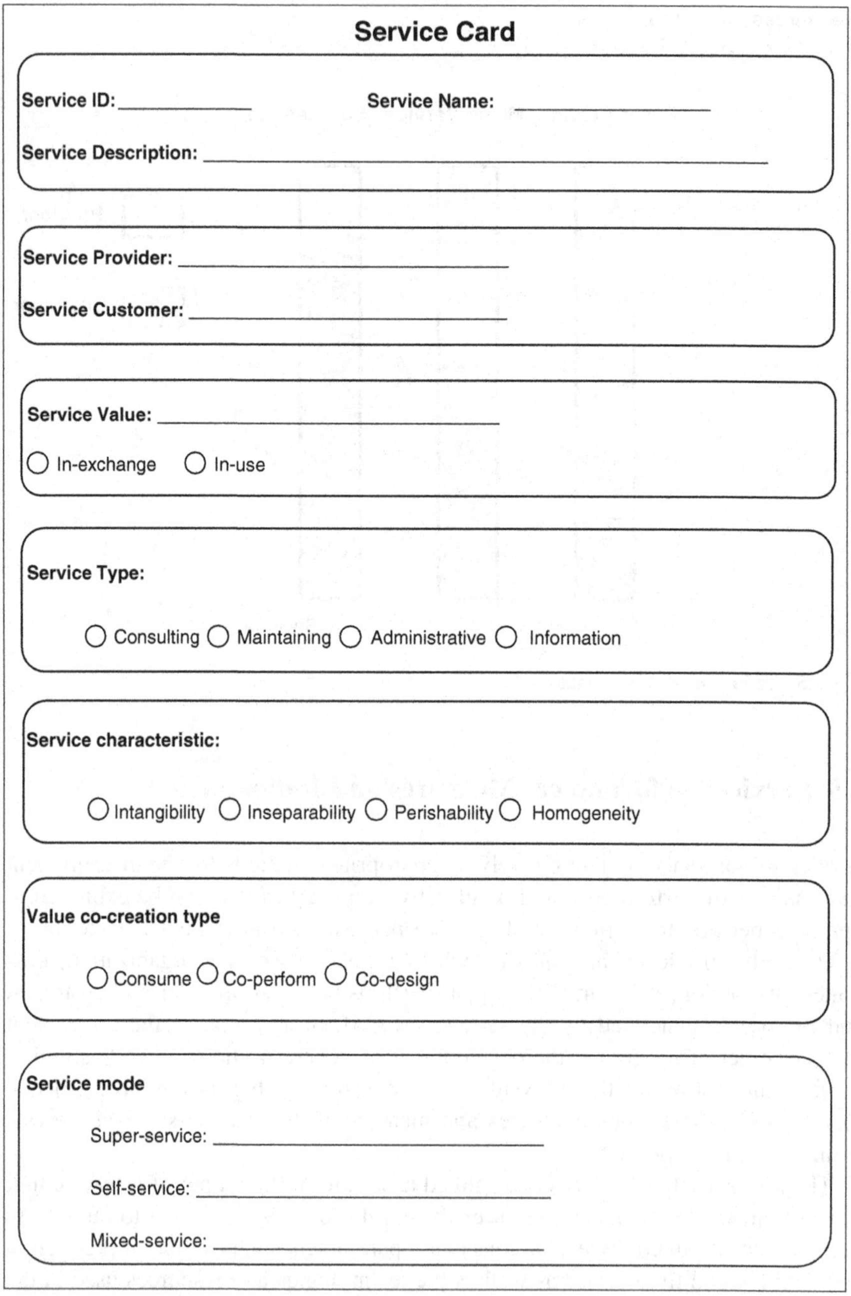

Service Card

Service ID:_____   Service Name: _____

Service Description: _____

Service Provider: _____

Service Customer: _____

Service Value: _____

◯ In-exchange   ◯ In-use

Service Type:

◯ Consulting ◯ Maintaining ◯ Administrative ◯ Information

Service characteristic:

◯ Intangibility ◯ Inseparability ◯ Perishability ◯ Homogeneity

Value co-creation type

◯ Consume ◯ Co-perform ◯ Co-design

Service mode

Super-service: _____

Self-service: _____

Mixed-service: _____

Fig. 2.6 Service card

use of the provider, service quality is determined mainly by the customers who compare what they expect the service to provide with what they actually receive.

2.10 The Service Sector

In general, the economy is based on the three sectors of *agriculture* and *manufacturing*, which produce tangible values, i.e. goods, and *services*, termed "the tertiary sector of the economy", which deliver intangible values. Although agriculture, which primarily supplies food, has been the basis of life and economics since antiquity, the industrial revolution shifted the center of mass of economics to the production of goods. However, in 2006, for the first time in history, the service sector employed more people globally (40.0 %) than either the agricultural (38.7 %) or manufacturing (21.3 %) sectors [36]. In addition, today the service sector also represents the largest sector in industrialized countries, and it is constantly growing (Fig. 2.7).

Finally, the continuing shift of the world economy toward services coupled with rapid advances in technology have significantly changed the way that organizations create, deliver and consume products, which consist of goods and services. At the same time, driven in large part by advances in technology, the service systems that are being designed to deliver services are becoming more comprehensive, complex and interdisciplinary [2, 38].

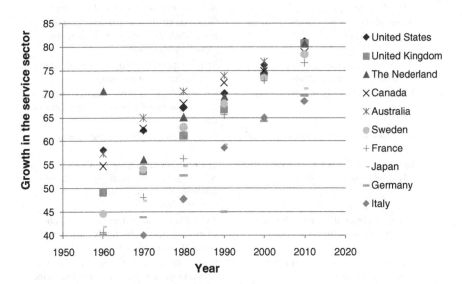

Fig. 2.7 Growth in the service sector (1960–2010) [52] for ten industrialized countries (percent GDP [37])

2.11 Service-Dominant-Logic

In 2004 Vargo and Lusch offered a new paradigm for thinking about commerce, marketing, and exchange known as *service-dominant* (*S-D*) *logic* [39–42]. They suggested that taking an S-D logic perspective would overcome many of the limitations inherent in the *goods-dominant* (*G-D*) *logic* that evolved during the Industrial Revolution. Moreover, S-D logic suggests that the focus should be moved from the exchange of products or goods to that of services, and that actually all exchanges between producers and consumers are based on service.

Vargo and Lusch also suggested ten foundational premises (FPs) of S-D logic (Table 2.2) [41]. In addition, S-D logic handles the basic constructs of social and economic exchange in a fundamentally different manner (Table 2.3) [42].

In addition, the S-D logic model suggests that the customer also participate in the production of the value, i.e., value in-use (Fig. 2.4b), and that the value is always co-created, jointly and reciprocally, in interactions among providers and customers using a constellation of integrated resources and capabilities that provider and customer share, combine and renew [21].

Table 2.2 Foundational premises (FPs) of service-dominant logic [41]

FPs	Premise	Explanation/justification
1	Service is the fundamental basis of exchange	The application of operant resources (knowledge and skills), "service" is the basis for all exchange. Service is exchanged for service
2	Indirect exchange masks the fundamental basis of exchange	Goods, money, and institutions mask the service-for-service nature of exchange
3	Goods are distribution mechanisms for service provision	Goods (both durable and non-durable) derive their value through use, i.e., the service they provide
4	Operant resources are the fundamental source of competitive advantage	The comparative ability to cause desired change drives competition
5	All economies are service economies	Service (singular) is only now becoming more apparent with increased specialization and outsourcing
6	The customer is always a co-creator of value	Implies value creation is interactional
7	The enterprise cannot deliver value, but only offer value propositions	The firm can offer its applied resources and collaboratively (interactively) create value following acceptance, but cannot create/deliver value alone
8	A service-centered view is inherently customer oriented and relational	Service is customer-determined and co-created; thus, it is inherently customer oriented and relational
9	All economic and social actors are resource integrators	Implies the context of value creation is in networks of networks (resource-integrators)
10	Value is always uniquely and phenomenologically determined by the beneficiary	Value is idiosyncratic, experiential, contextual, and meaning-laden

Table 2.3 Comparison of the G-D logic and S-D logic concepts [42]

Core constructs	G-D logic concepts	S-D logic concepts
Service	Goods and services	Serving and experiencing
	Transaction	Relationship and collaboration
Value	Value-added	Value co-creation
	Value-in-Exchange	Value-in-context
	Price	Value proposing
System	Supply-chain	Value-creation network
	Asymmetric information	Symmetric information flows
Interaction	Promotion/propaganda	Open source
	Maximizing behavior	Communication learning via exchange
Resources	Operand resources	Operant resources
	Resource acquisition	Resourcing

2.12   Service Science

As service philosophy has shifted from service products that are economically measurable and management-oriented to values that are more socially responsible and operationally-oriented, service research has become multidisciplinary, incorporating concepts from marketing, computer science, information systems, and operations as well as psychology and sociology. Hence, at about the same time that S-D logic was introduced, the IBM Almaden Research Center foresaw a need for a new discipline, which it named *service science*, with the goal of creating a platform for systematic service innovation [43–46]. The rationale and justification for this new discipline was to develop the underlying principles that define the subject matter and identify all the relevant stakeholder groups (e.g., academia, industry, and government) and demonstrate its relationship to other disciplines [47, 48]. Furtheron, the concept of S-D logic was synchronized with service science [49, 50].

Finally, one of the main incentives behind establishing a new track for service research in the form of service science was to clearly define service according to what it *is* and not, as was the convention in the past, according to what it is *not*, e.g., the service sector has traditionally been defined as whatever is not agriculture or manufacturing [51]. Service science, by necessity, takes a broad service system perspective that extends beyond the typical company or organization boundaries. Therefore, the goals in introducing this new discipline were (i) to recognize the importance of transdisciplinary research and teaching in contrast to the traditional functional disciplines, (ii) to apply scientific methods to better understand and manage services, and (iii) to develop valid metrics to measure performance.

References

1. Alter S (2008) Service system fundamentals: work system, value-chain, and life cycle. IBM Syst J 47:71–85
2. Fitzsimmons JA, Fitzsimmons MJ (2006) Service management: operations, strategy, and information technology, 5th edn. McGraw-Hill, Boston
3. Zeithaml VA, Parasuraman A, Leonard LB (1985) Problems and strategies in services marketing. J Mark 49(2):33–46
4. Edvardsson B, Gustafsson A, Roos I (2005) Service portraits in service research: a critical review. Int J Serv Ind Manage 16(1):107–121
5. Basole RC, Rouse WB (2008) Complexity of service value networks: conceptualization and empirical investigation. IBM Syst J 47:53–70
6. Peilin W (2008) The discussion of the research on service science. Libr J 27(3):2–8
7. Bitner MJ, Fisk RP, Brown SW (1993) Tracking the evolution of the services marketing literature. J Retail 69(1):61–103
8. Hartman DE, Lindgren JH Jr (1993) Consumer evaluations of goods and services: implications for services marketing. J Serv Marketing 7(2):4–15
9. Wolak R, Kalafatis S, Harris P (1998) An investigation into four characteristics of services. J Empir Gen Mark Sci 3(2):22–43
10. Regan WJ (1963) The service revolution. J Mark 47:57–62
11. Rathmell JM (1966) What is meant by services? J Mark 30:32–36
12. Shostack G (1977) Breaking free from product marketing. J Mark 41:73–80
13. Potts GW (1988) Exploit your product's service life cycle. Harv Bus Rev 66(5):32–36
14. Kuosa T, Westerlund L (2012) Service design-on the evolution of design expertise. Lahti Univ Appl Sci Series A Res Rep 17
15. Papazoglou MP, Georgakapoulos G (2003) Introduction to the special issue about service-oriented computing. Commun ACM 46(10):24–29
16. Arsanjani A (2004) Service-oriented modeling and architecture. IBM developer works
17. Brown A, Johnston SK, Larsen G, Palistrant J (2005) SOA development using the IBM rational software development platform: a practical guide. Rational Software 14
18. Bell M (2008) Service-oriented modeling: service analysis, design, and architecture. Wiley, New York
19. Demirkan H, Kauffman RJ, Vayghan JA, Fill H-G, Karagiannis D, Maglio PP (2008) Service-oriented technology and management: perspectives on research and practice for the coming decade. Electron Commer Res Appl 7(4):356–376
20. Krcmar H (2010) Informations management. Springer, Berlin
21. Vargo SL, Maglio PP, Akaka MA (2008) On value and value co-creation: a service systems and service logic perspective. Eur Manage J 26:145–152
22. Kuusisto A, Päällysaho S (2008) Customer role in service production and innovation-looking for directions for future research. Lappeenranta University of Technology, Report 195
23. Keaveney SM (1995) Customer switching behavior in service industries: an exploratory study. J Mark 59(2):71–82
24. Bitner MJ (2001) Service and technology: opportunities and paradoxes. Manag Serv Qual 11(6):375–379
25. Froehle CM (2006) Service personnel, technology, and their interaction in influencing customer satisfaction. Decis Sci 37(1):5–38
26. Glushko RJ (2010) Seven contexts for service system design. In: Maglio PP, Kieliszewski CA, Spohrer JC (eds) Handbook of service science. Springer, New York, pp 219–248
27. Meuter ML, Ostrom AL, Roundtree RI, Bitner MJ (2000) Self-service technologies: understanding customer satisfaction with technology-based service encounters. J Mark 64(3):50–64
28. Lin J-SC, Chang H-C (2011) The role of technology readiness in self-service technology acceptance. Manag Serv Qual 21(4):424–444

29. Campbell CS, Maglio PP, Davis M (2011) From self-service to super-service: how to shift the boundary between customer and provider. Inf Syst E-Bus Manage 9:173–191
30. Wolfson A, Tavor D, Mark S (2012) Sustainability and shifting from a 'person to person' to a super- or self-service. Int J u- and e- Serv Sci Technol 5(1):25–34
31. Balci B, Hollmann A, Rosenkranz C (2001) Service productivity: a literature review and research agenda. In: Proceedings of the XXI international RESER conference, Hamburg
32. Grönroos C, Ojasalo K (2004) Service productivity: towards a conceptualization of the transformation of inputs into economic results in services. J Bus Res 57(4):414–423
33. Johnston R, Jones P (2004) Service productivity: towards understanding the relationship between operational and customer productivity. Int J Product Perform Manage 53(3):201–213
34. Berry LL, Parasuraman A, Zeithaml VA (1988) The service-quality puzzle. Bus Horiz 31(5):35–43
35. Rust RT, Oliver RL (1994) Service quality: new directions in theory and practice. Sage Publications, Los-Angeles
36. Labour Office (2007) Global employment trends brief, Geneva, Jan 2007
37. U.S. Bureau of Labor Statistics (2011) Charting international comparisons of annual labor force statistics, Washington DC, Apr 2011
38. Karmakar U (2004) Will you survive the service evolution? Harv Bus Rev 82:100–107
39. Vargo SL, Lusch R (2004) Evolving to a new dominant logic for marketing. J Mark 68:1–17
40. Lusch RF, Vargo SL (2006) The service-dominant logic of marketing: dialog, debate, and directions. M.E. Sharpe Inc., New York
41. Vargo SL, Lusch RF (2008) Service-dominant logic: continuing the evolution. J Acad Mark Sci 36(1):1–10
42. Vargo SL, Lusch RF, Akaka MA (2010) Advancing services science with service-dominant logic: classification and conceptual development. In: Maglio PP, Kieliszewski CA, Spohrer JC (eds) Handbook of service science. Springer, New York, pp 132–156
43. Vargo SL, Maglio PP, Akaka MA (2008) On value and value co-creation: a service systems and service logic perspective. Eur Manage J 26:145–152
44. IBM Almaden Services Research (2006) Service science, management, and engineering (SSME): challenges, frameworks, and call for participation. IBM Almaden Research Center
45. IBM Almaden Services Research (2006) Service science, management, and engineering (SSME): what are services? IBM Almaden Research Center
46. Maglio PP (2007) Service science, management, and engineering (SSME): an interdisciplinary approach to service innovation. IBM Almaden Research Center
47. Maglio PP, Spohrer JC (2007) Fundamentals of service science. IBM Almaden Research Center
48. Katzan H (2008) Foundations of service science concepts and facilities. J Serv Sci 1(1):1–22
49. Maglio PP, Kieliszewski CA, Spohrer JC (2010) Handbook of service science. Springer, Berlin
50. Vargo SL, Akaka A (2009) Service-dominant logic as a foundation for service science: clarifications. Serv Sci 1(1):32–41
51. Lusch RF, Vargo SL, Wessels G (2008) Toward a conceptual foundation for service science: contributions from service-dominant logic. IBM Syst J 47(1):5–14
52. Wolfson A, Tavor D, Mark S, Schermann M, Krcmaar H (2010) S3-Sustainability and services science: novel perspective and challenge. Serv Sci 2(4):216–224

Chapter 3
Sustainability and Service

Abstract The service sector has undergone explosive growth in recent years, and today it is the largest sector of the economy. One channel through which the goal of sustainability can be achieved, therefore, is through the design of more sustainable services. In general, there are several different routes to imbue services with sustainability, from the rational use of resources to more efficient value co-creation processes to propositions of the same solution in an alternative, sustainable manner. Ecosystem services, which are nature's services that support and maintain life on earth, can be mimicked by a variety of service types and modes. These include environmental services that specialize in the minimization of environmental damage, green services that promote more efficient resource use and smaller environmental impacts, and eco-efficient services that are marketable systems of products and services capable of fulfilling a user's demand more sustainably. All service types and modes can be gathered under the umbrella of clean services (CleanServs), i.e., services that are competitive with, if not superior to, their conventional tangible or intangible counterparts and that reduce the use of natural resources and cut or eliminate emissions and wastes while increasing the responsibilities of both provider and customer. Finally, a sustainable service—which imbues the service's core-value with sustainability but that also requires the customer to become a provider of sustainability to current and future generations via the production and delivery of sustainable super-value—provides a framework for sustainability-based service innovation.

3.1 Why the Service Sector?

Although both our knowledge and awareness of environmental risk and sustainable development have increased significantly in recent years, the specific application of theory in practice is highly challenging. One channel through which this

© The Author(s) 2015
A. Wolfson et al., *Sustainability through Service*, SpringerBriefs in Applied Sciences and Technology, DOI 10.1007/978-3-319-12964-8_3

requirement can be addressed is to design more sustainable processes (in both manufacturing and services) that satisfy the needs of today's customers without sacrificing the ability of future generations to meet their own needs.

The combination of services—which represent the largest and most well-developed sector of the economy and the globally dominant employment sector—together with the introduction of new paradigms that promote viewing everything in terms of services, i.e., the service-logic dominant approach, suggests that the focus on sustainability should also include services. Hence, sustainability should be incorporated into services via all three main phases of the service life-cycle, i.e., design, production and delivery, and use. In addition, the various models and types of services that have been developed can be generally categorized into those that include sustainability as a key feature versus those in which the aspects of sustainability are of secondary importance or are even by-products of the process.

3.2 Ecosystem Services

Nature provides many services—from food and energy to air cleaning and temperature control by trees or water filtration by swamps—that promote and sustain ecosystems and that directly or indirectly benefit humanity, i.e., *ecosystem services* (Fig. 3.1).

The concept of ecosystem services refers to both tangible and intangible natural products that provide outputs and outcomes that directly and indirectly affect human wellbeing. In general, ecosystem services comprise four categories: provisioning, regulating, cultural, and supporting services [1, 2] (Table 3.1).

3.3 Environmental Services

The acute increases the world has witnessed in its population and in people's quality of life are associated not only with the implementation of mass production and the consumption of natural resources, but also with the discharge of effluents and waste. Meanwhile, people's knowledge about and awareness of the environmental risks associated with rising population densities have led to the inevitable

Fig. 3.1 Ecosystem services

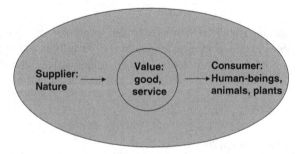

Table 3.1 Categories and examples of ecosystem services

Type	Description	Examples
Provisioning services	Tangible products obtained from ecosystems	Water, food, bio-chemicals, fuels, fibers
Regulating services	Intangible products obtained from the regulation of ecosystems	Water purification, air purification, water regulation, climate regulation, pollination
Cultural services	Intangible products obtained from ecosystems	Spiritual and religious value, social relations, aesthetic values
Supporting services	Intangible products required for the production and maintenance of other ecosystem services	Primary production, soil formation, nutrient cycling, provision of habitat

conclusion that their actions influence the environment for better or for worse. Indeed, the environmental risks that loom over humanity constitute one of the most urgent and complex problems facing the citizens of Earth.

Defined as the potential threat to the environment manifested in climate change and in land, water and air pollution, environmental risks are increasingly forcing governments, industries, and in fact every citizen of the world to find solutions by adopting novel ways of thinking and of behaving [3]. The environmental industry—which comprises technologies, products and services that benefit the environment and, by extension, also human-beings, by limiting the environmental impact of industrialization and urbanization—constitutes one such channel to mitigate environmental risks. According to the Organization for Economic Co-operation and Development (OECD), "the environmental industry consists of activities which produce goods and services to measure, prevent, limit and minimize or correct environmental damage to water, air and soil, as well as problems related to waste, noise, and eco-systems" [4].

The environmental industry grew tremendously in the final decades of the last century, accounting for a global market of $450 billion in 2000 that is on a par with pharmaceutical industry statistics from the same period [5]. Moreover, within the first decade of the 21st century, the global market for environmental goods and services grew by approximately 75 % to $782 billion by 2008 [6]. About 80 % of that market is divided between the US (38 %), Western Europe (28 %) and Japan (13 %). In the US, revenue from the environmental industry is broken down into the various industry segments of services (47 %), equipment (21 %), and resources (32 %).

As part of environmental industry market, *environmental services*—which comprise various combinations of scientific, technical, management and advisory activities specializing in the minimization of natural resource use and environmental damage, i.e., the treatment of post-use outcomes—also constitute growing markets in both developed and undeveloped countries (Fig. 3.2) [7]. As is the case for ecosystem services, the target of environmental services is sustainability, which is already introduced in the service design phase.

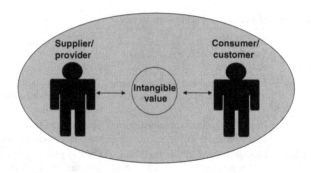

Fig. 3.2 Environmental services

In general, environmental services can be divided into three basic types:

(a) *Disposal services* provide an absorptive sink for residuals, e.g., the disposal, recycling, reclamation and re-use of waste and wastewater treatment.
(b) *Productive services* provide a rational use of natural resources, e.g., services that promote energy conservation projects.
(c) *Consumer or consumption services* provide for the physiological, recreational and other similar needs of human beings, e.g., environmental consulting, monitoring, and control.

Table 3.2 summarizes the US environmental services market by sector in 2008.

Finally, some of these services can be delivered solely by the supplier where the customer only consumes the value, for example, wastewater treatment. Others are delivered via co-creation, such that the supplier becomes a provider of the platform and the consumer participates in the delivery of the service and becomes a customer, e.g., a plastic bottle recycling service that is co-created by customers who assume the task of responsibly disposing of their bottles through a recycling facility service by first separating the bottles and then depositing each type of bottle in its designated bin.

Table 3.2 US environmental services market by sector in 2008 [8]

Sector	Billions ($)	Share (%)	Export (%)
Solid waste management	53.1	37	0.3
Hazardous waste management	9.2	6	1.0
Consulting and engineering	27.1	19	12.8
Remediation/industrial services	12.5	9	6.0
Analytical services	1.9	1	7.4
Water treatment works	40.7	28	0.6

3.4 Environmentally Friendly or Green Services

Since the 1970s people's awareness of the importance of reducing environmental pollution and of preserving Earth's natural resources has steadily grown, which has led to increased numbers and types of environmental regulations and policies. The beginning of the 21st century, however, saw a change in focus from a conventional 'end-of-pipe' treatment approach to using dedicated equipment and technologies to actively promoting the prevention of pollution and the rational use of natural resources by implementing policies based on product substitution and process modifications. In addition, under the general umbrella of designing environmentally friendlier processes, adopting corporate social responsibility, and realizing savings in costs, all of which contribute to improving the triple bottom line, sustainability practices are also translated into the production and delivery of both tangible and intangible products. In addition, various dimensions of sustainability have also become part of company strategic planning. Finally, sustainability initiatives can also serve as sources of customer and employee satisfaction and as indicators of quality for customers that transcend the immediate impact on profitability.

The above-mentioned change in focus has therefore stimulated the production and delivery of *environmentally friendly services* or *green services*, i.e., services that incorporate and promote more efficient resource use and reduced environmental impacts (Fig. 3.3). As in the case of tangible goods, here the emphasis is mainly on the resource based life-cycle of the service. Hence, the environmental

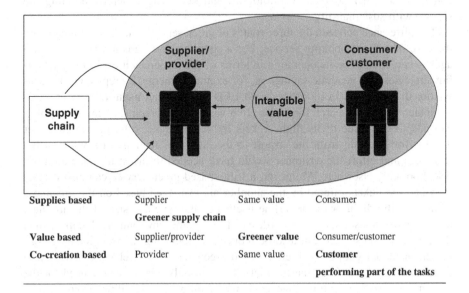

Supplies based	Supplier	Same value	Consumer
	Greener supply chain		
Value based	Supplier/provider	**Greener value**	Consumer/customer
Co-creation based	Provider	Same value	**Customer**
			performing part of the tasks

Fig. 3.3 Environmentally-friendly or green services

profit of environmentally friendly services is in fact defined mainly by the resources that are used to produce and deliver the essence of the value, and not by the value itself, and as such, it can also be defined as a 'by-product' of the process. Three different routes, based on the three main service components of provider, value, and customer, have been applied to achieve this goal: (i) *supplies-based*—making existing service production and delivery greener from the perspective of the supplier and his resources, i.e., the focus is on delivering the same value in a more environmentally sound way by adopting different approaches to the organization and management of the supplier's resources; (ii) *value-based*—greening the value itself by adding complementary features, where the focus is on creating new, environmentally conscious services; and (iii) *co-creation-based*—greening the service by reassigning the responsibilities for the resources, tasks and capabilities required in the creation of the service between the provider and the customer to realize a more efficient co-creation process. Nevertheless, although environmental concerns constitute the primary motivation behind the creation of environmentally friendly services, the main driver behind the greening of a service remains economic profit. As such, one's financial bottom line can be improved through the more effective and efficient use of the resources—e.g., materials and energy, manpower, and technologies—needed to create and implement the service. In addition, branding the service as green and attracting new clients can also advance its financial bottom line. Finally, new technologies, especially the Internet, have led to a revolution in our behavior and modes of communication as well as in how we perceive and implement services. Ultimately, this leads to the creation of more sustainable opportunities, such as the self-service mode, which usually enable more efficient and conscious use of resources, technologies, and knowledge, thereby imbuing the service with sustainability.

The difference between the three routes of green service can be illustrated using the case of a flight booking service. For a supplies-based green service, a travel agency can streamline its operation in terms of the resources it exploits by adopting strategies that generate savings in, for example, energy, by replacing the conventional light bulbs in its offices with LED lighting and using a more efficient computer system to search for flights. In a value-based green service, the supplier can add to the value of its flight tickets by, for example, offering the customer a ticket for the train from the airport to the city, thus realizing a savings in the resources and effort the customer would have needed to invest to make an additional booking elsewhere. Moreover, a value-based green service can also be that in which the tickets offered by the supplier are evaluated based on the greenness of the specific flight or of the airline itself, i.e., the type and size of the airplane, the company's sustainability record, etc. Finally, the environmental impact of a flight reservation service can also be reduced by replacing the traditional person-to-person model with a co-creation-based green service model via a self-service ticket booking mode (e.g., the Internet). This not only eliminates the need for the customer to drive to the travel agency, it also requires fewer facilities from the provider's side.

3.5 Products-Service Systems (PSS)

Generally speaking, both products and services comprise tangible and intangible values that are delivered from a supplier to a client to satisfy the demands of the latter. Because every product usually entails an element of service and every service is likewise typically based on the use of products, services and products usually exploited in combination as parts of a larger, loosely coordinated system that is focused on the fulfillment of a demand. Thus was born the concept of product-service systems (PSS) [9–12], and similar models such as integrated product service engineering (IPSE) [13, 14] and product service engineering (PSE) [15], during the last 15 years (Fig. 3.4). PSS was conceptualized as the addition of a service to a (new or an existing) product to obtain a combination with maximum added value, thereby producing a synergistic solution characterized by increased profitability and competitiveness. It was joined by another new concept, termed *servicizing*, which referred to the intensification of the service component of a PSS and that became synonymous with PSS [16]. With the aim of extending the efficiency and value of a good, servicizing dictates that the supplier provide functionality rather than a product and that the consumer, while not necessarily assuming ownership of the product, actually share with the supplier in the co-creation of the value. In contrast to green services, however, the product in PSS is part of the value, and together with the service, it supplies the customer with a solution.

With the increase in popularity of PSS as witnessed by the growth in research and development dedicated to its wider implementation, it has accrued myriad definitions, each of which takes a slightly different perspective on the framework of the system as well as on its aims and goals. One of the first and most straightforward definitions of PSS, "Marketable systems of products and services capable of fulfilling a user's demand", was proposed by Goedkoop et al. [17]. Hockerts and Weaver addressed the mixture of tangibility and intangibility inherent in PPS and claimed: "A pure product system is one in which all property rights are transferred from the product provider to the client on the point of sale. A pure service system is one in which all property rights remain with the service provider, and the clients obtain no other right besides consuming the service. A product service system is a mixture of the above. It requires that property rights remain distributed

Fig. 3.4 Product-service system (eco-efficient service)

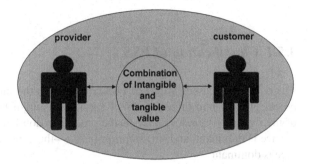

between client and provider, requiring more or less interaction over the life time of the PSS" [18]. At last, Manzini and Vezzoli [19] referred to PSS in terms of a business strategy: "An innovation strategy, shifting the business focus from designing and selling physical products only, to designing and selling a system of products and services which are jointly capable of fulfilling specific client demands".

Concomitant with the growth in interest in PSS due to their greater efficiency and profitability was a more focused consideration of the environmental impact of systems, a natural development owing to the opportunity inherent in the PSS concept to reduce the tangible elements, e.g., materials and energy, of system solutions. Revised definitions of PSS were subsequently offered that included an environmental or sustainability dimension with the aim of minimizing the system's environmental impact while maximizing its added value. These efforts also led to changes in the terminology, and PSS was referred to as *eco-efficient services* (EES), a concept that stresses the servicizing of the product and the shares in the responsibilities of the service as well as those of provider and customer in generating value [20–22].

EES is generally defined as "systems of products and services which are developed to cause a minimum environmental impact with a maximum added value" [23]. Alternatively, Baines et al. defined PSS (EES) as "an integrated product and service offering that delivers value in use. A PSS (EES) offers the opportunity to decouple economic success from material consumption and hence reduce the environmental impact of economic activity" [24]. Brandstötter et al. also added that "PSS (EES) tries to reach the goals of sustainable development", i.e., to seek a balance between the environmental, economic, and social dimensions of potential solutions [25].

In general, most researchers and designers perceive PSS or EES as a competitive proposal intended to satisfy consumer/customer demand. Therefore, the PSS (EES) model encompasses three main elements [12]: (i) the value is a combination of products and services, (ii) the value co-creation process is emphasized, and (iii) the solution is imbued with sustainability features, i.e., reducing resource utilization while increasing the responsibility of both supplier and customer and growing the economical profit. In addition to these three elements, the main characteristics of eco-efficiency include (i) increasing service intensity, (ii) reducing material and energy use, (iii) reducing waste and pollution discharge, (iv) extending PSS lifetimes, and (v) incorporating life-cycle principles. Yet as with the emphasis of green services, the main focus of PSS (EES) is materialization. In addition, it is resource-oriented, which refers to the production and use phases of the life-cycles of both the product and the service.

3.5.1 Classification of PSS

The numerous classifications of PSS contained in the literature [12] can generally be categorized into three groups of services distinguished by the dominant logic of their value: (i) *product-based*—the product is dominant, (ii) s*ervice-based*—the service is dominant, and (iii) *solution-based*—the combination of product and service is dominant.

Mathieu suggested two different types of PSS: (i) services supporting the product and that mainly assist the provider, and (ii) services supporting the actions of the customer and related to the co-creation process. Similarly, Manzini and Vezzoli also focus on what PSS offer, distinguishing between three types of service based on product ownership and use and how the decision-making power is distributed between the provider and the customer: (i) services that add value to the product, (ii) services that enable a platform for the customer, and (iii) services that provide a result to the customer [26]. Alternatively, Tukker offered a classification of PSS based on their orientation that includes three main categories: (i) *product-oriented service*—the product is the focus of the solution and owned by the customer, and it is delivered together with some extra, related service, (ii) *use-oriented service*—the product is still the focus of the solution, but it is owned by the provider, and it is used in a service mode by the customer, and (iii) *result-oriented service*—where provider and customer agree on a result and on the part played by the product in the delivery of this result [27]. Tukker suggested that on the PSS continuum between '*pure product*' and '*pure service*', eight types of PSS can be identified based on their economic and environmental characteristics:

1. **Product-oriented service**

    (a) *Product-related service*—the provider delivers to the customer a product together with services that are needed during the use-phase of the product.
    (b) *Advice and consultancy*—the provider delivers to the customer a product together with services related to the use of the product.

2. **Use-oriented services**

    (a) *Product lease*—the provider delivers to the customer a service that includes a product owned by the provider and used solely and for a limited time by the customer together with supportive and complementary services.
    (b) *Product renting or sharing*—the provider delivers to the customer a service that includes a product owned by the provider and used by the customer for a certain time or used sequentially with other customers together with supportive and complementary services.
    (c) *Product pooling*—the provider delivers to the customer a service that includes a product owned by one customer and used by other customers for a certain time, together with supportive and complementary services.

3. **Result-oriented services**

    (a) *Activity management/outsourcing*—the provider delivers to the customer a solution that comprises services and products and the customer does not buy the product but only the result of the product.
    (b) *Pay per service*—the provider delivers to the customer a solution that comprises services and products and the customer does not buy the product but only the result of the product according to the level of use.
    (c) *Functional result*—the provider delivers to the customer a solution that comprises services and products and how the solution is delivered is at the provider's sole discretion.

Table 3.3 Classification of PSS types

Class	Dominant	Product owner	Product user	Product decision-maker	Co-creation level
Product-oriented	Product	Customer	Customer	Customer	Low
Use-oriented	Service	Provider	Customer	Customer	Medium
Result-oriented	Product-service	Provider	Provider-Customer	Provider-Customer	High

The main elements of the various PSS types are summarized in Table 3.3.

Finally, though economic profit and efficiency were traditionally the primary incentives behind the design and development of PSS or EES early on, the effect of PSS on sustainability was also identified and recognized, and in many cases it was the main goal of the system. The difference between the various types of PSS can be illustrated using the transportation sector. While buying a private car is a 'pure product' solution, it is usually associated with product-related services such as maintenance of the car and consulting services such as those provided by an insurance company. Alternatively, the same solution can be realized by using a car leasing service, where the car, although owned by the provider, is delivered to the customer who also receives complementary services. The similar car rental model describes a service in which the customer usually receives the car for a more limited block of time and with fewer complementary services. In addition, multiple customers can group together and use a car-sharing model, in which they use a single car that is supplied by the provider for a specific journey. Likewise, they can use a car-pooling model where, to reach a common destination, they share (usually on a regular basis) a car owned by one of the members of their group. Finally, transportation services can also be result-based, such as those offered by an outsourcing transportation company, by a pay-per service in which a customer pays for use of the car based on amount of time and distance driven, or even by using public transportation, which is a functional result mode.

3.6 Clean Service (CleanServ)

As discussed from the beginning of this chapter, the various service models and types that have been offered over the years have aimed to fulfill customer demand while realizing the highest economic, social and environmental profits (Table 3.4). The primary difference between the types of services is in the main goal of each, but the service types are also distinguished by the composition of the value, i.e., tangible, intangible or both, and by the roles played by both the supplier/provider and the consumer/customer in the production and delivery of the service, i.e., value co-creation.

The different service models and modes can be divided generally into services that deliver only intangible value, i.e., 'pure services', where resources and goods

Table 3.4 Summary of various models and types of service

Service type	Definition	Supplier/provider	Value	Consumer/customer
Ecosystem services				
1. Provisioning	Naturals services that promote and sustain ecosystems	Nature	Serv./pro.	Consumer
2. Regulating				
3. Cultural				
4. Supporting				
Environmental services				
1. Disposal	Services specialized in the minimization of natural resource use and environmental damage	Supplier	Serv.	Consumer/customer
2. Productive				
3. Consumption				
Green services				
1. Supplies-based	Services that are more efficient with regard to resource use and environmental impact	Supplier	Serv.	Consumer
2. Value-based		Provider		Customer
3. Co-creation-based		Provider		Customer
Product-service systems				
1. Product-based (Product-oriented)	Marketable systems of products and services capable of fulfilling user demand	Supplier	Pro./serv.	Customer
(i) Product-relate				
(ii) Advice, consultancy				
2. Service-based (Use-oriented)		Provider	Pro./serv.	Consumer
(i) Leasing				
(ii) Renting or sharing				
(iii) Pooling				
3. Solution-based (Result-oriented)		Provider	Pro./serv.	Customer
(i) Outsourcing				
(ii) Pay per service				
(iii) Functional result				

are indirectly used to produce and deliver the value as a 'by-product', and services that deliver a combination of tangible and intangible value, i.e., PSS, where the product is directly delivered as part of the value. From the sustainability perspective, the essence of some 'pure services', like environmental services and in part ecosystem services, is the delivery of sustainability itself, yet any 'pure service' can be produced and delivered more sustainably, i.e., green service. On the other hand, PSS aim to gain added value by combining economic and environmental profits. Finally, with respect to imbuing service with sustainability, it is commonly accepted that the combination of rational resource use and the value co-creation process, which involves both the provider and the customer, has true potential to yield a more sustainable solution.

Because the production of goods is inherently tied to natural resource use and associated with the discharge of various by-products while services are essentially intangible, some have suggested that the goal of sustainability can be realized by delivering solutions that are based exclusively or mainly on services, i.e., *clean services—CleanServs* (Fig. 3.5) [28, 29]. A CleanServ is a service that is competitive with, if not superior to, its conventional tangible or intangible counterparts and one that reduces the use of natural resources and cuts or eliminates emissions and wastes while increasing the responsibilities of both provider and customer. As such, the concept of CleanServ covers the entire spectrum of services, from 'pure service' to PSS. But in the context of CleanServs, the service is at the core of the solution, i.e. *service-product systems*(SPS), expanding the definition of service and the connectivity between service and sustainability.

CleanServs comprise five categories defined in descending order from most to least sustainable, an assessment that is based on the service's 'clean character', related in part to its resource use. In addition, each category can also be

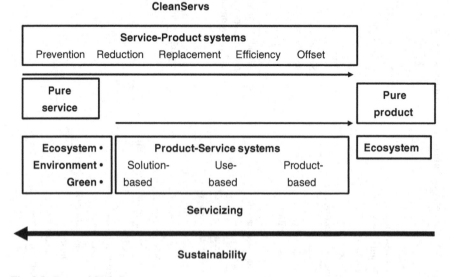

Fig. 3.5 Types of CleanServs

divided into sub-categories of service-orientation, e.g., self-service, e-service, etc. Note that the services that were defined above (Table 3.4) are included in these categories.

1. *Prevention*—the service offers an alternative solution to and prevents the production of a certain good or the delivery of another service. This solution is solely based on service delivery.
2. *Reduction*—the service offers an alternative solution based on the less intensive use of the same resources and technologies, i.e., same product or service. Thus, it increases sustainability by reducing the utilization of natural resources or the production of certain goods, thereby mitigating the corresponding environmental damage. This solution includes supplies-based and co-creation-based green services and product-based and service-based PSS.
3. *Replacement*—the service exploits other resources or technologies to deliver the same solution in the form of an alternative product or service, while increasing the sustainability and reducing the utilization of natural resources or the production of certain goods, thereby mitigating the corresponding environmental damage. This solution includes value-based green services and solution-based PSS.
4. *Efficiency*—the service improves operational performance and efficiency while reducing the consumption of materials and energy and the generation of waste and pollution in the production of a certain good or service. This solution is based on complementary services that are delivered in parallel during the production and delivery process and includes ecosystem services and environment services.
5. *Offset*—the service is complementary to other services by compensating for resource utilization during the production or delivery of goods or services. The service is thus delivered in parallel with the delivery of a certain product or service.

To illustrate the differences between the categories of CleanServs, the example of the transportation sector is used again, but it is expanded to include PSS. Moreover, the focus in the case of CleanServs is not on the product or the service component, but on the solution. For example, consider the variety of ways one can pay a bill. Many people would choose to take a trip to their local post office or bank, which can be done using different kinds of products, such as a car or a bike or using public transportation instead. The need to make the trip to the bank, however, can be eliminated by using an online system to pay the bill or a system that charges the bank account on a monthly basis or a pay by use service. Alternatively, that special car trip to the bank can also be avoided by combining several tasks in a single trip or using public transportation, in the process reducing private car use instead. Finally, the use of one's private car or the bill payment service can be run more efficiently, while the resources utilized in the process can be compensated for and offset by a service that offers to plant trees or to support certain environmental or social organizations, either monetarily or otherwise, at a price tag equivalent to that of the carbon footprint associated with the process [30].

3.7 Sustainable Services

As summarized above, there are various opportunities to incorporate sustainability into services. In general, it can be done either directly via the value that is delivered from the provider to the customer with or without co-creation process, or indirectly via changing or re-organizing the service's supply-chain as well as by delivering a combination of product and service. Yet the value-chain, which initiates with the provider and his supplies via the production and delivery of the value to the customer, should not end with the customer, as the purchase and use of a service by the customer also affects the natural and social environments and thus overall global sustainability. Hence, by using a service the customer also indirectly affects nature and other people, such as the current and even future generations. Thus, as a matter of fact, sustainability should first be an integral part of the underlying relationship between customer and service provider. As such, it should also account for all resources and effort that are incorporated into the direct value, i.e., the *core-value*. In addition, it should also be associated with the indirect delivery of resources and effort throughout the supply-chain as well as with the relationship between customer and other stakeholders, i.e., the 'by-products' as previously defined or the *super-value* (Fig. 3.6) [31, 32].

Because the sustainability of any service depends on both its core- and super-values, a service that sustainably delivers both values is defined as a *sustainable service*. Thus, a sustainable service is not only that in which sustainability is delivered via the core-value co-creation process while encouraging the supplier to rely on sustainable supplies during the production and delivery of the core-value. In addition, it also requires the customer to become a provider of sustainability via the production and delivery of sustainable super-value to current and future generations. Moreover, a sustainable service is also designed with potential long-term

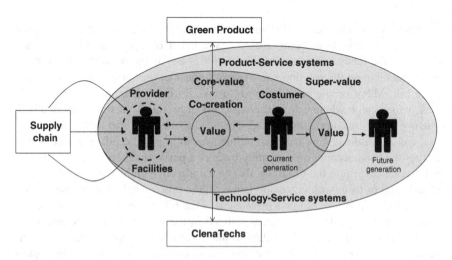

Fig. 3.6 Sustainable service

effects in mind, and it offers a novel perspective on the relationships between consumers, providers, and suppliers while fulfilling customer demands without negatively affecting the natural and social environments. Finally, a sustainable service also integrates tangible and intangible resources to create a life-cycle oriented scheme. With this broad definition of sustainable service, the customer becomes the main stakeholder responsible for the sustainability of the service, from deciding on whether to buy or use a service—by examining the service supply-chain and value-value—to considering the effects of the production, delivery and use of the service on future generations. Thus, this definition adds to the service multiple dimensions that include time and place as well as other customers.

Finally, sustainable service is a novel approach and a model that conceptualizes sustainability as a service while exploring the benefit of doing so, both for sustainability and for service science. In addition, sustainable service is also a framework for generating new and alternative sustainable values for supplying the same customer demands more sustainably, while considering the complexity of service [31]. As illustrated in Fig. 3.7, the model comprises two main stages: (a) a sustainable decision based on an evaluation of the service's resources, technologies, and information and knowledge, and (b) selection of the most sustainable choice from among the alternatives after evaluating each in terms of its integration of services and of manufacturing and agricultural processes, i.e., *Sustainability as Service Science—S³*. [31]. This new model encourages both providers and customers to adopt more environmentally conscious approaches while jointly designing and co-creating the values.

3.7.1 Nature Mimicry Approach

Nature has inspired and challenged humans since the beginning of life on Earth. Ecosystem services constitute an ideal example of sustainable, efficient, and compatible services that demonstrate a model for the design, production and delivery

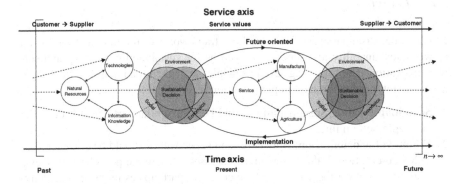

Fig. 3.7 S³-sustainability as service science model [31]

of sustainable service using a natural mimicry approach. In general, the natural mimicry approach is predicated on five fundamental prerequisites: it should be (i) future-oriented, it should consume (ii) minimum energy and involve (iii) maximum entropy, and it should be (iv) renewable and (v) evolutionary [33]. These five fundamentals are explained in detail as follows:

(i) *Sustaining life-future-oriented service*: As the driving force in nature is the perpetuation of life, each natural process or service is inherently future-oriented. Hence, sustainable services should incorporate future-oriented thinking to maintain global sustainability via service's super-value.

(ii) *Energy-service quantification*: Likewise, energy is the driving force of all natural processes, which are based on a minimum use of resources, i.e., material and energy, to sustain life. Sustainable services should therefore also be based on a minimum use of resources in the production and delivery of both the core- and super-values.

(iii) *Entropy-service quality*: Entropy dictates both the direction of a process as well as the efficiency with which resources are utilized. As such, it is also a measure of the number of ways in which a process can be produced. Thus, sustainable services should also be based on the most efficient routes for their production and delivery.

(iv) *Life-cycle-service renewability*: Nature works in cyclic processes. As such, sustainable services should be imbued with life-cycle perspectives and create renewable values that transfer sustainability from provider to customer, the latter of whom subsequently becomes a supplier of sustainability to future generations.

(v) *Evolutionary-spiral service*: Nature functions in an evolutionary fashion: populations adapt to changes and undergo selection. Likewise, sustainable services should be comprehensive, integrative, and smart, in the process monitoring changes in resources, technologies, and information and knowledge while incorporating those changes into all service life-cycle phases.

3.7.2 Principles

To conclude, sustainable service provides a framework for the mutual relationship between sustainability and services as well as for sustainability practice and service innovation. The following seven principles provide a road map for the design, development, production, and implementation of sustainable services:

1. **Sustainable value**: Sustainability should be an essential part of value delivery, i.e., triple bottom line.
2. **Whole-value orientation**: Both core- and super-values should be considered.
3. **Value co-creation**: Value co-creation processes—where the provider furnishes a platform to deliver the service and the customer participates in the service delivery by sharing some of the resources, tasks and capabilities—should be included.

4. **Value continuity**: The customer should be allowed and even compelled to be a provider of sustainability, i.e., from customer to provider.
5. **Value life-cycle**: All life-cycle phases and the resource-based life-cycle of the service should incorporate sustainability.
6. **Intangibility of value**: A solution should be delivered that is based on an intangible instead of a tangible product, i.e., prevent, reduce, replace, make efficient, or offset the use of goods.
7. **Sustainability as value**: Sustainability should be delivered as a service in and of itself.

References

1. Gretchen CD (1997) Nature's services, societal dependence on natural ecosystems. Island Press, Washington DC
2. http://www.millenniumassessment.org/en/index.html. Accessed 25 May 2014
3. Calow PP (1998) Handbook of environmental risk assessment and management. Wiley, New Jersey
4. The Environmental Goods and Services Industry (1999) Manual for data collection and analysis, OECD
5. International Trade Centre (2001) International trade forum, issue 2
6. Environmental Business International (2008) San Diego. CA, USA
7. Engel S, Pagiola S, Wunder S (2008) Designing payments for environmental services in theory and practice: an overview of the issues. Ecol Econ 65(4):663–674
8. Glossary of Environment Statistics (1997) Studies in methods, series F, No. 67, United Nations, New York
9. Sakao T, Lindahl L (2009) Introduction to product/service-system design. Springer, Berlin
10. Beuren FH, Ferreira MGG, Miguel PAC (2013) Product-service systems: a literature review on integrated products and services. J Clean Prod 47:222–231
11. Tukker A (2013) Product services for a resource-efficient and circular economy—a review. J Clean Prod. doi:10.1016/j.jclepro.2013.11.049
12. Gaiardelli P, Resta B, Martinez V, Pinto R, Albores P (2014) A classification model for product-service offerings. J Clean Prod 66:507–519
13. Lingegård S, Sakao T, Lindahl M (2012) Integrated product service engineering-factors influencing environmental performance. In: Matsumoto M, Umeda Y, Masui K, Fukushige S (eds) Design for innovative value towards a sustainable society. Springer, Dordrecht
14. Sundin E, Lindahl M, Comstock M, Shimomura Y, Sakao T (2007) Integrated product service engineering enabling mass customization. In: Proceedings of 19th International Conference on Production Research
15. Sakao T, Shimomura Y (2007) Service engineering: a novel engineering discipline for producers to increase value combining service and product. J Clean Prod 15(6):590–604
16. White AL, Stoughton M, Feng L (1999) Servicizing: the quiet transition to extended product responsibility. Report submitted to U.S. Environmental Protection Agency, Office of Solid Waste
17. Goedkoop MJ, van Halen JG, te Riele H, Rommens PJM (1999) Product service systems, ecological and economic basics. Ministry of Environment, The Hague
18. Hockerts K, Weaver N (2002) Are service systems worth our interest? Assessing the eco-efficiency of sustainable service systems. Working Document INSEAD, Fontainebleau
19. Manzini E, Vezolli C (2003) A strategic design approach to develop sustainable product service systems: examples taken from the 'environmentally friendly innovation' Italian prize. J Clean Prod 11(8):851–857

20. Hockerts K (1999) Eco-efficient service innovation: increasing business-ecological efficiency of products and services. In: Charter M (ed) Greener marketing: a global perspective on greener marketing practice. Greenleaf Publishing, Sheffield, pp 95–108
21. Bartolomeo M, dal Maso D, de Jong P, Eder P, Groenewegen P, Hopkinson P, James P, Nijhuis L, Örninge M, Scholl G, Slob A, Zaring O (2003) Eco-efficient producer services—what are they, how do they benefit customers and the environment and how likely are they to develop and be extensively utilized? J Clean Prod 11(8):829–837
22. Zaring O, Bartolomeo M, Eder P, Hopkinson P, Groenewegen P, James P, Örninge M (2001) Creating eco-efficient producer services. Gothenburg Research Institute, Gothenburg
23. Brezet JC, Bijma AS, Ehrenfeld J, Silvester S (2001) The design of eco-efficient services. TU Delft for the Dutch Ministry of Environment, Delft
24. Baines TS, Lightfoot HW, Evans S, Neely A, Greenough R, Peppard J, Roy R, Shehab E, Braganza A, Tiwari A, Alcock JR, Angus JP, Bastl M, Cousens A, Irving P, Johnson M, Kingston J, Lockett H, Martinez V, Michele P, Tranfield D, Walton IM, Wilson H (2007) State-of-the-art in product-service systems. Proc Inst Mech Eng Part B J Eng Manuf 221(10):1543–1552
25. Brandstötter M, Haberl M, Knoth R, Kopacek B, Kopacek P (2003) IT on demand towards an environmental conscious service system for Vienna. In: Proceedings of EcoDesign 03, third international symposium on environmentally conscious design and inverse manufacturing, Japan, pp 799–802
26. Mathieu V (2001) Service strategies within the manufacturing sector: benefits, costs and partnership. Int J Serv Ind Manage 12(5):451–475
27. Tukker A (2004) Eight types of product-service system: eight ways to sustainability? Experiences from SusProNet. Bus Strategy Environ 13(4):246–260
28. Wolfson A, Tavor D, Mark S (2013) From CleanTech to CleanServ. Serv Sci 5(1):193–196
29. Wolfson A, Tavor D, Mark S (2014) CleanServs-clean services for a more sustainable world. Serv Sci Sustain Acc Manage Policy J 5(4):405–424
30. For example: http://www.trees4lifecampaign.com/reducing-your-carbon-footprint/. Accessed 25 May 2014
31. Wolfson A, Tavor D, Mark S, Schermann M, Krcmaar H (2010) S^3-sustainability and services science: novel perspective and challenge. Serv Sci 2(4):216–224
32. Wolfson A, Tavor D, Mark S (2013) Sustainbilility as service. Sustain Acc Manage Policy J 4(1):103–114
33. Wolfson A, Tavor D, Mark S (2011) Sustainable service: the natural mimicry approach. J Serv Sci Manage 4:125–131

Chapter 4
Assessing Sustainable Services

Abstract The design, production and delivery of sustainable services as well as the ability to assess and compare different services to identify those that are the most sustainable, requires that the interactions between sustainability and service be somehow 'engineered'. This goal can be achieved by instituting a sustainability assessment for services based on a set of unified and comparative measures to qualify and quantify the sustainability of services with respect to their core- and super-values and in terms of the service co-creation process. In this chapter, we suggest a methodology, with indicators and indexes, to measure the integration of sustainability in the design of the service supply chain and to assess the sustainability of service. Taken together, the use of such indicators promotes more efficient decision-making regarding service sustainability. In addition, they enable services and processes to be compared from the perspective of their supply-chains, a process that subsequently helps identify the most un-sustainable links in those chains as well as the links that have the greatest impact on the sustainability of the entire service system. Finally, they also allow for the incorporation of complementary and supportive services to increase the sustainability of the service system as a whole.

4.1 Decision-Making Process

The consideration of environmental issues and their integration—together with the social and economic dimensions of society—into sustainable development have become critical in decision-making processes. However, despite our familiarity with the concept of sustainability, which has become a commonly used word in our daily jargon and has even crossed the gap from vision to action, the sustainability of a product, a service or a process is still difficult to precisely define and accurately measure. More importantly, accurately assessing the effects our activities and decisions have on global sustainability remains a formidable challenge.

© The Author(s) 2015

A. Wolfson et al., *Sustainability through Service*, SpringerBriefs in Applied Sciences and Technology, DOI 10.1007/978-3-319-12964-8_4

We all make decisions on a daily basis. Some of our decisions, like when to wake up in the morning or what to eat, are quick and simple, and they are made instinctively or on a logical or an emotional basis. Other, more complex decisions, which entail more interdisciplinary features and which may potentially affect our lives in the long term and influence other people, require more organized decision-making. The process of decision-making involves identifying and choosing between alternatives to reach certain goals [1, 2]. It is based on the gathering and subsequent conversion of information into knowledge to reduce uncertainty regarding the differences between alternatives, among which a reasonable choice can be made. In the development context, decision-making involves the design of strategies, the definition of policies, and the execution of actions.

A variety of decision-making methodologies can be used to complete an effective decision-making process. In general, the stages of a decision-making process begin with a definition of the situation and of the goal, i.e., the desired outcome, of each step in the process and of the process as a whole. The second phase entails generating alternatives based on the decision criteria that have been identified, which usually depends on the knowledge, experience, and skills of those involved. Then the third stage involves information gathering and processing to enable decision makers to consciously select between the alternatives in the final stage, after which they can put it into action. As such, decision-making models should facilitate good judgments. The most commonly used model lists the pros and cons of each alternative and has the ability to assign each a different weight, based on its importance and on accepted policy, and finally, the highest scored option can be selected.

As stated above, the decision-making process is usually based on valid information and knowledge that are normally translated into measures that can be used to represent each situation, thereby facilitating a comparative assessment of the alternatives. Although there exist different levels of measurement and different measuring tools, all measurement methods seek to quantify a situation and represent it with an absolute or relative number. Thus, the first step in assessment is organizing information into representative numbers or indicators that should be as clear and simple as possible, understandable and easily calculated and used, and available and reliable. However, as each indicator is limited and narrow in scope, multiple sources of data are then aggregated into indexes that also account for the interaction of different indicators and that present broader perspectives. Note that the selection of appropriate indicators and indexes is crucial and that not everything should be measured or can be monitored.

4.2 Assessment of Service's Sustainability

Regarding the comprehensive, complex and multidisciplinary concept of sustainability, it is clear that whether large or small, any decision that purports to be sustainable should be made by thinking beyond short-term profit while accounting for the integration of environmental, social and economic elements. In addition, each decision should consider its effect not only on as broad a selection of people as possible, but also its effect on nature.

As discussed in Chaps. 1 and 2, the assessment and measurement of sustainability and service, respectively, are complicated tasks. In addition, although there is currently a variety of measures to quantify the level of sustainability of a process, most of them are based exclusively on the environmental aspects of sustainability, i.e., resource consumption, pollution emission and waste generation. Such measures are expressed, for example, by the ecological or carbon footprints of a service/process or by the environmental sustainability index of a country. In the case of services, on the other hand, the most commonly used measures are the service's productivity, efficiency and quality.

Assessments of both sustainability and service should integrate numerous factors and account not only for various fields and disciplines, but also for more general features, such as time, place and culture. Herein lies the difficulty of representing the two concepts with an absolute, simple and visible number. Moreover, the integration of the two results in interaction between variables and subsequently greater complexity. Therefore, an assessment system of sustainable services should begin with the design phase of the service life-cycle and continue with repeated evaluations during the production and delivery of the service. This can be achieved via two main parallel routes that comprise complementary levels of focus: (i) *macro-level*-developing measurements that consider the comparative sustainability of the whole service, and (ii) *micro-level*-developing methods to assess the sustainability of various entities in the service's supply-chain and the effects those entities have on the sustainability of the whole service system. Furthermore, to integrate sustainability orientation and service measures, a service should be assessed from the perspectives of four main categories: (i) resources, like materials and energy, (ii) facilities, for example, office and technologies, (iii) effort, such as time and labor, and (iv) knowledge and information. In addition, service also includes a co-creation process, while sustainable service comprises both core- and super-values together with the integration of products and services. Therefore, the sustainability assessment of a service should also consider the *co-creation level* (the division of resources, tasks and capabilities between the provider and the customer), its *core- to super-value ratio* (the division of the above-mentioned components between core- and super values), and its *product to service ratio* (the product elements vs. the service elements of a PSS). Yet although some studies have investigated theoretical and empirical methods for measuring co-creation [3–8] and PSS [7, 8], neither the level of co-creation of a service nor the share of a product in PSS is currently easy to define or measure. This is especially true when comparing services across industries and even in different locations. Finally, because the indicators and measures that have been proposed use different scales and units, the development of viable comparative measures that reflect the relative impact of a service on sustainability requires that the sustainability measures of service be normalized to a unified arbitrary unit.

4.3 Macro-level: Methodology, Model and Measures

Sustainability assessments of services at the macro-level, i.e., the whole system, usually adopt life-cycle analysis models such as is done with tangible products, the latter of which is mainly resource oriented. Thus, to assess the sustainability of the whole service-system, a *sustainability number* calculated by integrating environmental, economic, and social indicators is offered. The sustainability number should consider the type of balance that exists between the direct and indirect inputs of resources, such as water, energy, materials and manpower and investments, and the outcomes of the service, such as pollution emission and economic growth [9]. However, as the value of each indicator can vary widely, it is necessary to normalize all indicators to a unified and comparative scale. Therefore, the first step in evaluating the sustainability number is to calculate a normalized parameter, Pi, for each indicator i that reflects its weighted value. The second step of the assessment is to separately calculate the environmental, economic and social impacts—ENi, ECi and SOi, respectively—by combining the values of all normalized corresponding indicators into one index. In this step, one cannot neglect the fact that different parameters are often assigned different weights based on their relative importance in the process or their level of participation in delivering the service (Eq. 4.1).

$$ENi = f_1(Pi), \quad ECi = f_2(Pi), \quad SOi = f_3(Pi) \tag{4.1}$$

In the last step, the sustainability number of the whole service is calculated by aggregating the three normalized indexes, ENi, ECi, and SOi (Eq. 4.2).

$$Sustainability\ number = f(ENi, ECi, SOi) \tag{4.2}$$

Finally, the calculations and the assessment are done in terms of the division in responsibilities between the provider and the customer for the resources and tasks associated with the service. This division allows for the different elements of sustainability assigned to the provider and to the customer to be distinguished, thereby enabling calculation of the corresponding measure, the *sustainability level of co-creation* (Eq. 4.3). As illustrated in Eq. 4.3, the higher the sustainability level of co-creation, the greater the responsibility of the customer in supporting the sustainability of the service. Likewise, the *sustainability value-ratio* represents the split between the core-value and the super-value in terms of resources and tasks (Eq. 4.4), and as such, it enables one to determine quantities of resources that are assigned to the direct goal of the service versus those that are by-products.

$$Sustainability\ co-creation\ level = \frac{Sustainability\ number - customer}{Sustainability\ number - provider} \tag{4.3}$$

$$Sustainability\ value-ratio = \frac{Sustainability\ number - Core-value}{Sustainability\ number - Super-value} \tag{4.4}$$

4.3.1 Bee-Factor

As a decision-making tool, any sustainability measure should be clear and simple to calculate, but it should also reflect the influence, on the sustainability of a service or a process, of changes or investments in that process. This setup helps ensure that the right decision—which should also include an understanding of where more effort should be invested in the service chain—is made. But while current sustainability measures consider the global effect of a service or a process, they often lack tangible measures that would help decision makers understand the potential effects that certain changes could have on the whole system and that are critical to the decision-making process. Therefore, although they enable comparisons of alternatives, these comparisons cannot constitute true sustainability measures because they are unable to identify possibilities for improving a process, such as where to invest first in a given process to achieve the greatest improvement. Thus, the difficulties inherent in measuring absolute sustainability could be overcome by assessing the change in sustainability using a measure based on profit and loss. Such a method could then be exploited as a powerful decision-making tool to evaluate the effects that certain changes will have on the sustainability of a product or a process and to compare alternatives.

Cost-effectiveness or cost-benefit analyses, which generally entail calculating and comparing the costs and outcomes or benefits of a product or a process, are acceptable tools for managing and planning activities, and as such, they are highly applicable to sustainability assessment [10]. Insofar as nature supplies humanity with both products and services in the most effective and sustainable manner, a nature mimicry approach based on economic profit-loss analyses of investments should be adopted. An example from nature that can illustrate such a cost-benefit analysis is manifested in a comparison of all the resources required by a bee to maintain its own life with the benefit gained by humanity from the bee's flower pollination activities, which are essential to sustaining life on Earth as we know it. This measure can be translated into a ratio of the area of land that the bee's service maintains to that required to support its own life. Likewise, we suggest using a *bee-factor*, a comparative dimensionless number, i.e., sustainability difference to cost difference, to represent the influence that an action or a service has on global sustainability (Eq. 4.5). It resembles a ratio of the relative direct and indirect 'outcome' versus income entailed in a given action or service. Specifically, it expresses the relationship between the change in sustainability (in terms of profit or loss due to the implementation of a given decision) and the cost difference as defined by the investment or return relative to a certain benchmark. As such, it is applicable to comparisons of pairs and/or combinations of processes, products and services. Moreover, in certain cases where there is not enough data to calculate the absolute sustainability of a service/process, calculating the sustainability difference may be easier.

$$Bee-Factor = \frac{benefit}{cost} = \frac{outcome}{income}$$
$$= \frac{sustainability\ difference\ (profit\ or\ loss)}{|cost\ difference\ (relative\ to\ a\ benchmark)|} \quad (4.5)$$

In general, the sustainability difference is calculated by considering the change, between two states or alternatives, in the components of sustainability, i.e., environmental, economic, and social profit or loss, ΔENi, ΔECi and ΔSOi. To calculate the difference, the values of ENi, ECi and SOi at a given time are compared to those of a pre-defined benchmark (calculations done using Eq. 4.1). The outputs of these two calculations can then be translated into monetary price tags, and the difference between them will be either positive (profit) or negative (loss). Similarly, the cost difference can be calculated by considering the changes effected by actions that lead to either investment, which yields a negative difference, or to return, which yields a positive difference. But because sustainability is the main goal, the investment difference will always be positive, as defined by the absolute value sign in Eq. 4.5.

The bee-factor is thus a relative sustainability measure that facilitates comparisons not only between alternatives to the same solution, but also between identical services across places and cultures. Moreover, use of the bee-factor effectively translates sustainability into a tangible and accessible economic measure in the form of a monetary price tag that can be used on a daily basis, both as a policy determinant at the administrative level and by laypeople. Note that it is clear from Eq. 4.5 that the higher the bee-factor of a given product/service is, the greater the influence of global sustainability. Furthermore, as an inherently comparative measure, the bee-factor represents a change in sustainability compared to an existing state or the difference between two alternatives, and it refers to a product/service that is used in a particular time and place. Thus, it can vary from negative to positive corresponding, respectively, to sustainability differences that show a loss to

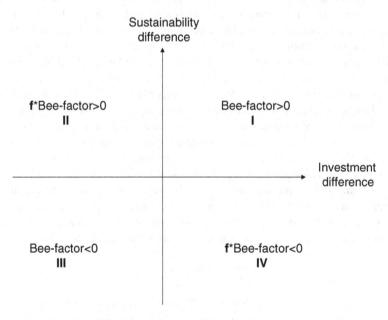

Fig. 4.1 Representation of the four possible outcomes of calculations based on the bee-factor

those that show a profit. Indeed, according to the definition of the bee-factor, when the sustainability difference is positive, regardless of the sign of the cost difference, the bee-factor is always positive, as can be represented in a Cartesian system (quadrants I and II, Fig. 4.1). Likewise, when the sustainability difference is negative, either a positive or a negative cost difference results in a negative value for the bee-factor (quadrants III and IV, Fig. 4.1). Based on this representation, the most sustainable option is when the sustainability difference is positive and the cost difference is actually negative, i.e., the change between the two routes leads to profit (quadrant II, Fig. 4.1). To distinguish it from the case when both the sustainability difference and the cost difference are positive (quadrant I, Fig. 4.1), a factor that is higher than unity should be assigned to the bee-factor in quadrant II. Likewise, when the sustainability difference is negative and the cost difference is positive (quadrant IV, Fig. 4.1), a scenario that reflects the least sustainable option, the bee-factor should also be assigned with a factor that is higher than unity.

4.4 Micro-level: Methodology, Model and Measures

By definition, sustainable service is that in which the core-value is conceived as a link in a chain that connects supplies with the end-customer in a manner that promotes sustainability as a super-value to future generations (Fig. 4.2). As such, service providers should ensure both the sustainability of their supplies and the ability of their customers to supply sustainability to the next generations. For their part, customers should also be involved in the design of the service core-value.

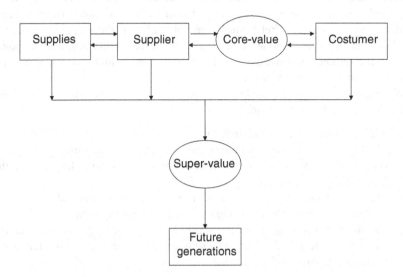

Fig. 4.2 Sustainable service supply-chain

Sustainable service thus accounts for the impacts of the provider and its supplies, on the one hand, and on the other hand, of the customer and how he/she uses the service vis-à-vis global sustainability. Moreover, the conceptualization of sustainable service links services with other services and processes. Therefore, sustainability assessments of services analyze the service supply-chain and all its entities and activities, which themselves are services.

Sustainability assessments of entire service supply-chains enable comparisons of different services and processes based on their supply-chains. In addition, the sustainability assessment is also a tool to identify the most non-sustainable links in the supply-chain as well as those with the greatest impact on the sustainability of the whole system. Thus, a methodology for the service sustainability assessment will comprise definitions of the building blocks of the service supply-chain and evaluations of their sustainability. It will also include the direct and indirect impacts of each building block on the sustainability of others, and it will assess the overall sustainability of the service. In addition, sustainability assessments can identify potential complementary and supported services that can be incorporated into the supply-chain to increase its overall sustainability.

In constructing the sustainability profile of each basic element of the supply-chain and its contribution to the overall sustainability of the service, the main mission of the service, i.e., propagation of the service core-value, is given greater emphasis in the assessment procedure. This means that if the service is, for example, supplying customers with fuel in a gasoline station, although the production and refining processes used to produce gasoline probably have more deleterious environmental impacts than are entailed in refilling one's vehicle with gasoline at the station, the focus of the sustainability assessment should be on the refueling service. In addition, the supplier (gasoline station) may be able to reduce its environmental impact, thereby increasing the sustainability of the service, by reducing water and electricity use at the gasoline station by, for example, using complementary, indirect services. Yet because the core-value of the service is refueling people's vehicles with gasoline, accordingly, the sustainability assessment should focus on the station's fuel storage and refueling systems, which are supporting services to the core-value of the service.

We suggest that the sustainability assessment of a service on a micro-level should pursue the following scheme:

1. Identify service type and its main actors.
   a. Identify the core-value, which includes identifying the direct participants (i.e., providers and customers), the solution that the service delivers, and the exact value of the service.
   b. Identify the super-value, i.e., indirect providers, customers and values, recognizing that sustainability is the main focus of the super-value.
   c. Identify the class (e.g. green service, PSS, etc.), type (e.g. product-oriented, value-base, etc.) and mode (e.g. self-service, super-service or traditional "person to person") of the service.

2. Map service-value-chain actions and entities, i.e., the service's building blocks.

3. Calculate:
   a. Quantity measures: the sustainability of each building block (*Sustainability number$_j$*) and the sustainability of the whole service (*Sustainability number*).
   b. Quality measures: the *Sustainability impact factor* and the *Sustainability ranking*.

4.4.1 Quantity Measures

We offer two measures to assess the sustainability of each building block and of the whole supply-chain of a service: the *sustainability number$_j$*, which corresponds to building block *j*, and the *sustainability number* of the whole service system, which is calculated by integrating environmental, economic, and social indicators.

Sustainability number$_j$ is calculated as described above for the sustainability number (Eqs. 4.6 and 4.7), where Pij represents normalized parameters for each indicator i and building block j, and ENij, ECij, and SOij represent the parameters normalized to the environmental, economic and social indexes, respectively.

$$ENij = f_1(Pij), \quad ECij = f_2(Pij), \quad SOij = f_3(Pij) \tag{4.6}$$

$$Sustainability\ number_j = f(ENij, ECij, SOij) \tag{4.7}$$

As a final step of the methodology, the building blocks are assembled into a number representative of the capacity of the whole service supply-chain to govern and deploy the entire service, i.e., *sustainability number*, which is derived from the sustainability number$_j$ of each basic building block (Eq. 4.8).

$$Sustainability\ number = f(sustainability\ number)_j \tag{4.8}$$

4.4.2 Quality Measures

In determining the effect of each individual building block of a service on the sustainability both of other building blocks and of the entire service supply-chain, measurements of sustainability quality facilitate the identification of methods for improving service sustainability. We therefore suggest two additional sustainability measures—*sustainability impact factor* and *sustainability ranking*—to assess the quality of sustainability.

In view of the inevitable effects that interactions between the different building blocks have on the sustainability of the whole service/process, the *sustainability impact factor* represents the impact of an individual building block on the sustainability of the whole service. It depends both on the number of building blocks that directly interact with a certain building block and on the *sustainability number$_j$* of each building block. As such, *the sustainability impact factor* of a given building block is calculated by dividing the average *sustainability number$_j$* of all downstream building blocks directly involved with it by the average *sustainability number$_j$* of those that are upstream.

As the service supply-chain comprises many entities that also have indirect impacts on each other, the *sustainability ranking* reflects not only the quantity, but also the quality, of the connections between various services in terms of the sustainability of the whole service system. The *sustainability ranking* of a given building block is calculated by dividing the average *sustainability number$_j$* of all downstream building blocks that directly and indirectly affect it by the average *sustainability number$_j$* of all such upstream building blocks. Based on a link analysis algorithm similar to that used by PageRank® [11], the measure facilitates the identification and ranking of the basic services that have the largest effects on the sustainability of the service system as a whole.

The measures of *sustainability impact factor* and *sustainability ranking* can be employed not only to identify the weakest links in the service supply-chain, but also to increase the chain's sustainability by adding supporting and complementary services during the design process. The latter task can be done by searching the services database for alternative sustainable basic services whose *sustainability numbers$_j$* are larger than those of the services that constitute the supply-chain design and by adding supporting services with high *sustainability impact factors*, thus improving the sustainability of the entire service system by increasing the *sustainability rankings* of its other services.

References

1. Klein GA, Orasanu J, Calderwood R (1993) Decision making in action: models and methods. Ablex Publishing, New York
2. Stone DA (1997) Policy paradox: the art of political decision making. WW Norton, New York
3. Rust RT, Lemon KN, Zeithaml VA (2004) Return on marketing: using customer equity to focus marketing strategy. J Mark 68:109–127
4. Reay P, Seddighi HR (2004) An empirical evaluation of management and operational capabilities for innovation via co-creation. Eur J Innov Manage 15:259–275
5. Field JM (2012) Designing service processes to unlock value. Business Expert Press, New York
6. Emmenegger MF, Frischknecht R, Stutz M, Guggisberg M, Witschi R, Otto T (2006) Life cycle assessment of the mobile communication system UMTS: towards eco-efficient systems. Int J Life Cycle Assess 11(4):265–276
7. Zabalza Bribián I, Capilla AV, Usón AA (2011) Life cycle assessment of building materials: Comparative analysis of energy and environmental impacts and evaluation of the eco-efficiency improvement potential. Build Environ 46(5):1133–1140
8. Wolfson A, Tavor D, Mark S, Schermann M, Krcmar H (2010) S3-Sustainability and services science: novel perspective and challenge. Serv Sci 2:216–224
9. Diaz-Balteiro L, Romero C (2004) In search of a natural systems sustainability index. Ecol Econ 49:401–405
10. Cellini SR, Kee JE (2010) Cost-effectiveness and cost-benefit analysis. In: Wholey JS, Hatry HP, Newcomer KE (eds) Handbook of practical program evaluation. Jossey-Bass a Wiley Imprint, San Francisco
11. Brin S, Page L (1998) The anatomy of a large-scale hypertextual web search engine. Comput Netw ISDN Syst 30:107–117

Chapter 5
Examples

Abstract Several examples of the interaction between service and sustainability are discussed. The division of resources, facilities, effort and knowledge between the provider and the customer as well as the core- and the super-values delivered are addressed. These examples provide insight about realistic opportunities to imbue services with sustainability and to design more sustainable solutions.

5.1 Introduction

In the previous chapters the concepts of sustainability and service and their mutual relationship were discussed and a methodology and measures to assess the sustainability of services and improve their performances and their impacts on the natural and social environments were offered. In this chapter, the sustainability of various services of different types and modes is discussed and illustrated with several examples. However, due to a lack of data about the sustainability of various services, the carbon footprint, which accounts for the emission of greenhouse gases during the production and delivery of a service or a product, is used in most cases as the representative sustainability indicator. In addition, if possible, the indicators are presented according to the methodology presented in Chap. 4 which is also the basis for calculations of the corresponding sustainability measures. As such, the four categories—resources, facilities, effort and knowledge and information— identified as important to sustainability assessments are used while the division of resources and tasks between provider and customer and between the core- and super-values offered by the service are also accounted for.

© The Author(s) 2015
A. Wolfson et al., *Sustainability through Service*, SpringerBriefs in Applied
Sciences and Technology, DOI 10.1007/978-3-319-12964-8_5

5.2 Gasoline Filling Station

Identity card

Provider   gasoline station
Customer   driver
Core-value   gasoline filling
Super-value   gasoline station and car operations
Service type   (1) pure-service, (2) co-creation-based green service
CleanServ class   (2) reduction
Service mode   (1) super-service, (2) self-service
Co-creation type (1) consume, (2) co-produce

In the gasoline filling station type of service, two service modes are offered: super-service, i.e., a full-service station, where the provider pumps the gas, and a self-service station, where the customer pumps the gas. Correspondingly, the filling station entails two possible types of co-creation: consume, in which the customer gets fuel from the station, or co-produce, in which the customer performs some of the service delivery tasks, e.g., the gasoline filling. Tables 5.1 and 5.2 illustrate the division of resources and capabilities between the provider and the customer in terms of the co-creation process and sustainability, respectively, in both service modes, and they also show the core-value (CV) and super-value (SV) of the service.

Regarding the co-creation process, it can be seen from Table 5.1 that both the resources and the facilities are identical for the two operation modes and are ascribed mainly to the service provider. On the other hand, the effort and knowledge required for gasoline filling are completely shifted from the provider in super-service to the customer in self-service. Table 5.2 shows a similar scenario for sustainability. In addition, to make the service greener, the provider can increase filling station sustainability by incorporating supplies-based services, e.g., improving resource management and utilization in the station, while customers can increase their impact on sustainability by choosing a more sustainable type of car, driving the car in a more

Table 5.1 Co-creation division of resources and capabilities in a gasoline filling station

	Value	1. Super-service		2. Self-service	
		Provider	Consumer	Provider	Customer
Resources	CV	Gasoline	Gasoline	Gasoline	Gasoline
	SV	Electricity, water, manpower, etc.	Car	Electricity, water, manpower, etc	Car
Facilities	CV	Pumps	None	Pumps	None
	SV	Station	Car	Station	Car
Effort	CV	Filling	None	None	Filling
	SV	Station operation	Car operation	Station operation	Car operation
Knowledge	CV	Filling	None	None	Filling
	SV	Station operation	Car operation	Station operation	Car operation

Table 5.2 Sustainability division of resources and capabilities in a gasoline filling station

	Value	1. Super-service		2. Self-service	
		Provider	Consumer	Provider	Customer
Resources	CV	Gasoline—land and water pollution	Gasoline—using more sustainably produced gasoline and additives	Gasoline—land and water pollution	Gasoline—using more sustainably produced gasoline and additives
	SV	Rational use of resources—electricity, water, paper etc., and treatment of wastewater, garbage, etc.	Air pollution	Rational use of resources—electricity, water, paper etc., and treatment of wastewater, garbage, etc.	Air pollution
Facilities	CV	Safety precautions—controlling gasoline emissions, underground gasoline leakage, etc.	Efficient car	Safety precautions—controlling gasoline emissions, underground gasoline leakage, etc.	Efficient car
	SV	Clean technologies	Car safety equipment	Clean technologies	Car safety equipment
Effort	CV	Sustainable filling	None	None	Sustainable filling
	SV	None	None	None	None
Knowledge	CV	Sustainable filling—signposting etc.	None	Sustainable filling—signposting etc.	None
	SV	Rational use of resources		Rational use of resources	Efficient car use

sustainable fashion, etc. However, these actions are related mainly to the super-value or the by-product of the service and not to its essence, i.e., gasoline filling.

From the perspective of sustainability, the gasoline filling operation of a filling station can also be improved by maintaining the filling system and by installing technology to prevent air pollution and the leakage of fuel into the ground. In addition, the service's sustainability can also be changed by redistributing resources and tasks between the provider and the customer, i.e., co-creation-based green service. Shifting the service tasks from the provider to the customer, thereby increasing the co-creation level, also redistributes the responsibility for sustainability between the two. It also means that while in a super-service framework, the provider is in charge of pumping the gasoline—conferring on the provider greater responsibility for the sustainability of the activity, e.g., loss of fuel to spillage—in the self-service scenario, the customer assumes that responsibility. For the co-creation process to be effective, however, the customers must be informed about how to fill their cars' gasoline tanks properly without having a negative impact on the environment. Regardless of customer knowledge, assigning some of the effort invested in performing the service to the customers allows the provider to save money on other station operational resources, e.g., paying employees. For their part, customers who pump their own fuel are much more aware of (and therefore, have more control over) the amounts of gasoline they use, thereby providing more sustainable super-value and increasing the sustainability of the entire service. To conclude, the increased sharing between provider and customer (i.e., their active participation) in resource management and use illustrated in this example shows that increasing the value co-creation level facilitates greater flexibility and awareness vis-à-vis sustainability. In addition, it creates more opportunities to increase the sustainability of the service.

5.3 Money Transactions

Identity card

Provider	bank
Customer	any bank customer
Core-value	money transactions
Super-value	bank and car or computer operation
Service type	(1) pure-service, (2) co-creation-based green service
CleanServ class	(2) replacement
Service mode	(1) super-service, (2) self-service
Co-creation type	(1) consume, (2) co-produce

A co-performed service, in which customers perform some of the service tasks using their own resources and capabilities, can be illustrated by an Internet based self-service, which replaces the conventional person-to-person service mode. One such service is e-banking, which allows bank customers to remotely conduct online financial activities, for example, financial transactions over a secure website operated and offered by the institution instead of driving to the bank to make the transaction through a bank teller. Table 5.3 illustrates the division, between provider and customer, of resources and capabilities in both the super- and self-service modes, while considering the service core-value (CV) and super-value (SV).

From Table 5.3 it can be seen that using e-banking services for money transactions decreases the resources and effort expended by both provider and customer, and as such, their respective investments and expenses decline while both the efficiency and sustainability of the service grow. However, both service modes can realize further increases in sustainability via potential actions that the provider, i.e., the bank, could take. For example, the bank could promote increased awareness of sustainability among, and encourage the active participation of, both its employees and its customers by adopting a green service model of the supplies- and co-creation based types, respectively. An assessment by HomeStreet Bank to measure its carbon footprint found that it could be easily reduced with several, easily implemented measures. For example, these could include adopting energy saving measures, cutting the use of paper, and/or reducing the number of times per week its employees need to commute to work. A reduction in employee commuting could be realized with compressed work week policies and by installing the necessary technical support that would allow employees to telecommute and participate in virtual meetings to reduce the bank's indirect transportation footprint. In addition, that transportation footprint could also be reduced by employee use of alternative transportation modes instead of private cars. These include bicycles, the use of which typically require that the employer install bike tracks, lockers and showers for its employees, and carpooling, among others [1].

Table 5.3 Comparison of the division between consumer/customer and provider of resources and capabilities in the super-service mode, i.e., traditional person to person, and in the self-service transaction

	Value	1. Super-service		2. Self-service	
		Provider	Consumer	Provider	Customer
Resources	CV	Paper, electricity	None	Electricity	Electricity
	SV	Electricity, water, manpower, etc.	Gasoline	Electricity, water, manpower, etc.	None
Facilities	CV	Bank branch	None	Website	Computer
	SV	Bank	Car	Bank	Home/office
Effort	CV	Manpower	None	Manpower	Internet use
	SV	Bank operation	Driving to the bank, standing in line, car operation	Bank operation	Bank operation
Knowledge	CV	Bank system operation	None	Bank system operation	Internet use
	SV	Bank operation	Car operation	Bank operation	No
Carbon footprint (g CO_2)-total		115[a]		7.3[b]	

[a]Calculated based on carbon footprint measurement per employee for a year by HomeStreet Bank [1], assuming 25 working days per month, 10 h workdays, and 5 min for each transaction

[b]Calculated based on annual CO_2 emission of 300 million tons and 1.9 billion internet users for 365 days/year, 10 h use at 10 min for a transaction (usually, the customer needs more time than the teller to complete the transaction)

5.4 Music Market

Identity card

Provider	music company
Customer	music fans
Core-value	music transaction
Super-value	music production and office, warehouse, car and computer operation
Service type	product-service system: (1) product-oriented service, product-related type, (2) use-oriented, product sharing type, (3) result-oriented, functional result type
CleanServ class	(2) reduction, (3) replacement
Service mode	(1) super-service, (2) mixed-service, (3) self-service
Co-creation type	(1) consume, (2) co-produce, (3). co-produce

The third example, in which sustainability is again expressed by carbon footprint, is illustrated with a music purchasing service. Three scenarios are compared: the traditional person-to-person retail service where the customer drives to a CD shop and buys a CD, an e-commerce service where the customer uses the Internet to purchase a CD that is delivered by the provider to the customer's home, and a digital music downloading service that also exploits the Internet (Fig. 5.1 and Table 5.4). The supply-chains of the three service systems were also compared by calculating the carbon footprint of each action/entity.

Because the resources and greenhouse gas emissions pertaining to the actual recording of the music are identical for the three different service delivery processes, they were excluded in this analysis. Likewise, the greenhouse gas emissions associated with the customer's home were excluded.

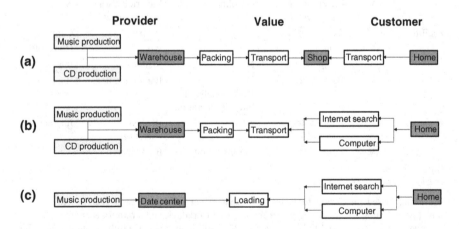

Fig. 5.1 Supply-chain of three music purchase scenarios. *White* action, *dark gray* facility, *light gray* production

Table 5.4 Division of co-creation and greenhouse gas emissions (gCO$_2$) in the purchase of music [2]

	Value	(1) Retail	(2) e-Commerce	(3) Digital
Transport	CV	1,850	675	0
Loading energy	CV	0	0	10
Internet search	CV	0	10	10
Computer energy	CV	0	10	10
Retail store energy	CV	250	0	0
CD production	SV	250	250	0
Packing	SV	575	875	0
Warehouse energy	SV	250	250	0
Date center energy	SV	0	0	400
Sum	CV + SV	3,175	2,070	430
Sustainability co-creation level		1.39	0.01	0.08
Sustainability value-ratio		1.95	0.51	0.08

The greenhouse gas emissions of the core-value (CV) and super-value (SV) components of the service are presented in Table 5.4, which shows that in the retail (Fig. 5.1a) and e-commerce (Fig. 5.1b) modes, CD production occurs prior to customer demand, thereby creating a need for a factory and a warehouse. Thus, the two modes only differ in terms of the resources, facilities, effort and knowledge invested in the CD delivery process. While for the retail mode, the customer assumes responsibility for the bulk of the requirements associated with CD delivery by driving to and from the retail location, using e-commerce divides the tasks and resources needed for the delivery between customer and provider. In the e-commerce scenario, however, the customers must have the facilities and knowledge and must invest the electricity and time to order the CD over the Internet, and product delivery from the warehouse to the customers' homes is done by the provider. Hence, choosing e-commerce over retail results in different divisions of resources and tasks between provider and customer and requires a higher level of co-creation. In addition, e-commerce also promotes streamlining of the service delivery process, as the provider can combine deliveries to multiple customers in the same journey, thereby reducing the amounts of greenhouse gases it emits per delivery.

Finally, the purchase of music via digital download (Fig. 5.1c) offers a new and much more sustainable alternative that is clearly illustrated by the reduced emissions of greenhouse gasses associated with this mode (Table 5.4). In this scenario, both provider and customer improve their performances by co-creating the service delivery process and efficiently dividing the resources and tasks between them. Here the provider no longer needs to manufacture a CD, pack it and physically deliver it to the customer, whose need to travel to a retail store is eliminated while sharing with the provider the resources, facilities, effort and knowledge that are required to deliver the service. Indeed, the digital download mode of music purchases significantly reduces the resources of both provider and customer, generating added-value and profit for both and increasing the sustainability of the service.

An examination of each scenario's sustainability co-creation level, which represents the division in resources and tasks between the customer and the provider, shows that increasing the co-creation level of this service increases its overall sustainability (Table 5.4). Yet the sustainability of the service does not depend only on the level of co-creation, but also on the quality of the co-creation process with respect to resource investment. For example, though the move from e-commerce to digital download reduces the overall carbon footprint, the sustainability co-creation level is increased by the greater share in relative energy use of the customer. In addition, the sustainability value ratio, which represents the ratio between the sustainability that was related to the core-value versus that due to the super-value, is decreased, which shows that greenhouse gas emission and thereby system sustainability is due mainly to the super-value. Thus, while the efficiency of the core-value, on the one hand, is significantly increased, there is still room, on the other hand, to improve the sustainability of the super-value.

From the perspective of an analysis based solely on carbon footprint, the bee-factor associated with a change from retail CD purchases to the e-commerce or digital modes is positive. In this case, a positive bee-factor is the result of the reductions in the carbon footprints (and thus, the increase in sustainability) of the two alternatives to the retail mode. Moreover, the value of the bee-factor is even higher for the digital purchase mode, in which the provider benefits from a reduction in its investments: delivery to its customers of the music they purchased requires neither warehouse facilities nor the resources or effort needed to produce, package or ship the CD to the customer.

5.5 Transport

Identity card

Provider	transportation company
Customer	passenger
Core-value	journey
Super-value	car production and maintenance
Service type	product-service system
CleanServ class	reduction, replacement, efficiency
Service mode	super-service, mixed-service, self-service
Co-creation type	consume, co-produce, co-design

One of the most widespread activities in our daily lives, the transportation of people is generally defined as the movement of individuals from one location to another. But transportation, which is necessary for both economic and social development, represents a sustainability dilemma. Inherently dependent on the use of vehicles, transportation is associated with heavy environmental pollution, both from the direct pollution caused by vehicle use and the indirect negative environmental impact associated with the production and maintenance of vehicles and their disposal at the end

Table 5.5 Core- and super-value components in an environmental assessment of passenger transportation [3]

	Car and bus	Light rail
Core-value		
Operational	Running, idling	Running, idling, auxiliaries
Super-value		
Non-operation-direct	Manufacturing, maintenance, insurance	Manufacturing, maintenance, insurance
Non-operation-indirect—Infrastructure	Roadway: construction, maintenance, lighting and parking	Station and track conservation, maintenance, control, cleaning and lighting
Non-operational-indirect—Fuels	Fuel production and distribution	Electricity generation

of their economic lives. As described in Chap. 3 for PSS, the gasoline-based private car, i.e., PSS of the product-oriented class and product-relate type, can be replaced by several alternative services and transportation solutions, such as car leasing, i.e., PSS of the service-oriented class and product lease service type, or car renting, i.e., PSS of the use-oriented class and product renting or sharing type. However, both these solutions do not really change the overall sustainability as both use the same vehicle, a car, with a private driver. Alternatively, there are also more sustainable PSS alternatives, like carpooling, i.e., PSS of the use-oriented class and product pooling type and CleanServs of the reduction class, or public transportation, i.e., PSS of the result-oriented class and pay per service type and CleanServs of the replacement class.

Chester and Horvath examined several transportation solutions and scenarios in their study of how passenger transportation is environmentally assessed [3]. They calculated the greenhouse gas emissions per person per kilometer traveled (PKT) in equivalent carbon dioxide for direct operation of the car, i.e., core-value, and for indirect, non-operational components, such as vehicle production and maintenance and the construction and maintenance of infrastructure like roads and parking, and insurance and fuel production, i.e., super-value (Table 5.5). Using the carbon foot-print as the primary measure of sustainability, they compared the use of a private car with one passenger to public transportation via a train or bus service and car-pooling, where several people share a private car (Fig. 5.2).

The results in Fig. 5.2 primarily illustrate that with the exception of light rail, in all scenarios, the emissions associated with the operational direct core-value are higher than those from the non-operational super-value. Furthermore, in general, a shift from a mode of transportation based on a private car with a single passenger to either public transportation or carpooling requires customer co-creation and reduces greenhouse gas emissions. As such, although public transportation requires some effort and knowledge from the customer, the main type of customer co-creation is consuming the service, which is related to the service core-value, i.e., movement from one place to another. This suggests that service sustainability is not only affected by the co-creation behavior of a certain customer, but also by the extent of involvement in the co-creation process of all the customers of that service. Moreover, the

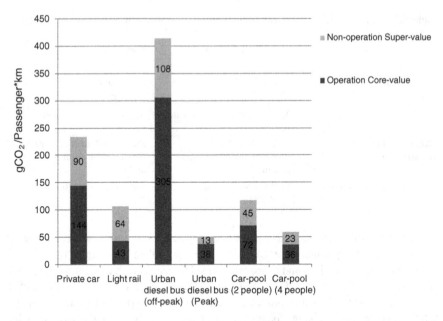

Fig. 5.2 Carbon footprints of some transportation solutions [3]

use of public transport such as a bus during off-peak hours can result in low numbers of passengers, which leads to carbon dioxide emissions per passenger that are even higher than those per passenger for private car journeys (Fig. 5.2).

In general, service sustainability can be increased by changing some of the resources or technologies used in the service. For example, the provider can use more environmentally friendly fuel or replace some of its buses with hybrid buses. For their part, the customers can increase service sustainability by using a mobile bus schedule and route planner application. Yet the sustainability of the service can also be increased by adopting a more efficient co-creation process, in which the provider adapts its resources and facilities to the customer's needs and habits. One way this could be accomplished is by using minibuses during off-peak hours, thereby enabling the provider to run its bus service more sustainably at full or close to full capacity.

Another service that can be used for transportation is carpooling. Customers of this service not only consume the service, they also partially co-produce it by performing some of the tasks, such as driving, filling gasoline, etc. To co-create the service core-value, however, carpooling services also require significantly more coordination between their users, who actually co-design the service by determining, for example, the time and the route that associated with public transportation. Obviously, moving from two- to four-person carpooling also decreases the greenhouse gas emission per person. In addition, the efficiency of the service co-design is manifested in the number of customers, which is higher the more efficient the service. Likewise, the schedule and route of the journey are planned both for the needs of the whole group and also to the benefit of each customer, thereby leading to increased service sustainability.

The most complex co-production pathway is co-design. From a sustainability point of view, the co-design of services—either from the innovation stage or to adapt existing services to changes to ensure that they continue to be sustainable—is usually beneficial, as it often results in more efficient service based on a rational use of resources and reduced discharge of pollutants to the environment. Two mobile applications, Waze [4] and Tiramisu [5], are examples of the real-time co-design of transportation market services that can increase the overall sustainability of transportation, i.e., they are CleanServs of the efficiency type. Waze, billed as the world's largest community-based traffic and navigation application, allows its customers to plan their routes, thereby saving both time and fuel. The second application is designed to track the transit system in real time. Described as a "crowd-powered transit information system" [5], Tiramisu is a service developed to improve user transit experiences and transit accessibility. As such, it enables the user to monitor bus arrivals in real time, including information about whether a given bus has any open seats, and to submit online reports to the bus company about any problems they may have encountered on the bus. In addition, the application also allows the driver to report as to whether the bus is running on schedule. A service like Tiramisu is expected to increase customer satisfaction with public transportation, with the ultimate aim being to increase the number of customers that use public bus services, i.e., consume.

The success of Tiramisu and similar services, however, also requires active customer participation, i.e., that customers co-perform some of the tasks, in the production and delivery of the service's value. In the case of Tiramisu, tasks the customers can co-perform include investing effort and knowledge while using their own facilities (home pc, mobile phone) to access the information posted by the Tiramisu service (to plan their bus journey) or to report the real-time data upon which the service depends, a process that eventually leads to the better design of the bus routes and schedule. Such ongoing co-design of the bus service, in turn, promotes greater efficiency on the part of the provider, while expanding the base of customers that regularly access Tiramisu for its online reporting of timetable changes increases the number of customers using the bus service. The ultimate result comprises increases both in the co-production level and in the service sustainability.

In addition to the carbon footprint, which measures the overall consumption of materials and energy associated with each transportation solution, all modes of modern transportation entail environmental impacts of varying degrees in the form of air pollution due to the emissions to the atmosphere of pollutants, such as SO_2, NOx, CO and volatile organic compounds (VOC) (Table 5.6). The emissions of these pollutants can be attributed to operational components, i.e., vehicle operation, or non-operational components, i.e., vehicle production. Likewise, the amounts emitted by each transportation solution vary, such that, for example, NOx emission is highest with buses, while the heaviest CO emissions are generated by private cars.

The data presented in Fig. 5.2 and Table 5.6 illustrate the difficulty in presenting a clear picture of transportation solution sustainability when the assessment is based in part on lists of the environmental indicators for each service. Because the average person will be challenged to make the best decision using such lists to compare the service options, there is a need for indexes, each of which combines several indicators into an easy to understand number. However, existing indexes

Table 5.6 Air pollutant emissions of some popular transportation solutions

		Private car	Light rail	Urban diesel bus (off-peak)	Urban diesel bus (peak)
SO$_2$ (mg/PKT)[*]	Operation core-value	8.14	238.36	2.72	0
	Non-operation super-value	204.04	265.83	228.75	28.6
NOx (mg/PKT)	Operation core-value	399.76	13.68	2474	309.25
	Non-operation super-value	263.04	150.86	256.24	32.03
CO (mg/PKT)	Operation core-value	7157.59	0	553.83	69.23
	Non-operation super-value	521.99	393.95	653.93	74.73
VOC (mg/PKT)	Operation core-value	261.13	3.77	68.48	8.56
	Non-operation super-value	297.63	85.89	289.85	36.24
Safety[**] [6]		0.62	0.06	0.03	

[*]Milligrams per passenger per kilometer traveled
[**]Transportation accident death rate, United States, 2006 average

for transportation that aggregate several air pollution indicators into a single index, such as the *air pollution index* or the *air quality index*, are currently only calculated at the city and country levels. In addition, the sustainability of each means of transportation also depends on other environmental factors, such as the land used in the operation of a certain transportation mode, on social aspects, such as safety (Table 5.6), and on economic aspects like the overall cost of the mode. These issues illustrate the difficulty inherent in measuring service quality and sustainability and of clearly showing their relationship.

5.6 Labeling

Identity card
Provider   labeling company
Customer   any customer
Core-value   decision-making comparative tool
Super-value   surveys, office, computer operation
Service type   pure-service
CleanServ class   efficiency
Service mode   mixed-service
Co-creation type co-design

An example of an efficiency class CleanServ is illustrated by services that rank products or services for the benefit of the customer, sometimes even labeling

Table 5.7 Sustainability rank of different shoe brands [7]

Brand	Ranking
Puma, Nike, Adidas, Reebok	B
Umbro, Helly Hansen	C
Champion, New balance, Asics	D
Saucony, Fila, Diadora	E

the product. Ranking or labeling is usually based, in part or in combination, on surveys of experts, provider information and client feedback. While this kind of service allows customers to choose their product or service, it also indirectly encourages the providers to improve their products or services by offering more efficient processes and complementary or supportive services.

The European brand comparison website Rank a Brand is an example of a consumer-run sustainability assessment service where over 500 product and service brands, including from the food and beverages, media, fashion, electronics, energy and travel sectors, are ranked to allow customers to make informed choices [7]. Table 5.7 lists the rankings of several shoe brands, for example, which are calculated based on environmental issues such as carbon emission or water use policies and on social issues such as labor conditions. Rank a Brand rankings are grades (from G to A) of increasing sustainability that allow customers to choose the shoe brands they want not merely based on brand popularity, performance and price, but also on the sustainability performance of the producer.

Similarly, carbon labeling—first introduced in 2006 by the Carbon Trust Company from the UK [8], a service that labels products with their carbon footprints—enables customers to compare products on a single scale based on the overall greenhouse gas emission of the product over its entire life-cycle. Product labeling with the carbon footprint not only enables consumers to choose the most sustainable product from among several alternatives, it also recruits them as co-creators of the service. Thus, the widespread adoption of carbon labeling and the subsequent increased consumer awareness can lead to changes in consumer buying behavior that can motivate the affected providers to strive for greater efficiency (and not only retain, but also add to their customer base). In so doing, they will deliver sustainability to the next generations as a super-value.

References

1. Seattle climate partnership, homeStreet bank—A carbon footprint measurement case study (2009). http://nbis.org/nbisresources/case_histories/homestreet_carbon%20footprint_case_study.pdf. Accessed 25 May 2014
2. Weber CL, Koomey JG, Matthews HS (2009) The energy and climate change impacts of different music delivery methods, Final report to Microsoft Corporation and Intel Corporation. http://download.intel.com/pressroom/pdf/cdsvsdownloadsrelease.pdf. Accessed 25 May 2014
3. Chester MV, Arpad H (2009) Environmental assessment of passenger transportation should include infrastructure and supply-chains. Environ Res Lett 4(2):024008

4. https://www.waze.com. Accessed 25 May 2014
5. http://www.tiramisutransit.com. Accessed 25 May 2014
6. U.S. Department of Transportation (2014) Bureau of transportation statistics, 2014
7. http://www.rankabrand.org. Accessed 25 May 2014
8. http://www.carbontrust.com. Accessed 25 May 2014

Chapter 6
Sustainability as a Service

Abstract Insofar as it propagates intangible value such as information, knowledge, awareness, methods, and tools, among others, sustainability is essentially a service. Yet it is inherently complex and necessarily comprehensive in nature. As such, it should account for and holistically integrate the environmental, social and economic dimensions of life. But it is not enough to merely account for those dimensions as whole entities, but rather, they should be broken down into and examined at various levels and scales, i.e., from individuals to societies, from the short- to the long-term, and from local to global. Such a task is understandably impossible to deliver in a single service, and therefore, sustainability as a service usually involves multiple providers and customers who must take active part in the value co-creation process to produce and deliver the service. Nevertheless, defining sustainability as a service extends the relations between sustainability and service beyond the incorporation of sustainability into services, directing the focus on sustainability itself. In addition, it allows sustainability practice to adopt the methodologies, frameworks, knowledge, methods and tools that were designed and developed to produce and deliver services and to organize the features of sustainability into a well-defined value-chain.

6.1 Definition

Although today, sustainability is already a well-known and widely accepted concept and paradigm [1, 2]—incorporated in many levels of the production and delivery of goods and services and their combinations and in decision-making processes—making the jump from theory to action is neither trivial nor straightforward. In addition, despite the generally more open attitude to the adoption of sustainable behavior, it is not always clear how local actions can effect global change or how current behavior will affect future generations. Moreover, despite the challenges associated with clearly defining the practice of sustainability, its application is still limited to cases where it is convenient, comfortable and cost effective.

© The Author(s) 2015   73
A. Wolfson et al., *Sustainability through Service*, SpringerBriefs in Applied
Sciences and Technology, DOI 10.1007/978-3-319-12964-8_6

As sustainability is immaterial, vague and complex, comprising the interests of multiple stakeholders [1, 2], it can be defined as an intangible value that is delivered from provider to customer via a co-creation process, i.e., a service. This definition is in agreement with the service-dominant logic concept, according to which all exchanges between provider and customer are based on services, even when resources and goods are involved [3–5]. It also conforms to the value-in-use model, which includes the co-creation process wherein provider and customer cooperate and share in the resources, tasks, capabilities and knowledge involved in the production and delivery of the service.

The main incentive behind defining sustainability as a service is to direct the focus on sustainability itself to ensure that sustainability is the primary driving force and core-value of the service, and not merely a by-product or connotative element of some other process or action. Furthermore, defining sustainability as a service dictates that the relations between sustainability and service are reciprocal and not limited to the incorporation of sustainability into services. In addition, it also enables the creation of an organized framework to facilitate the active implementation of sustainability. Such a framework should characterize the nature both of the value itself and of the roles played by the participants in the value co-creation process.

Linking sustainability with service—in the process defining the frameworks, knowledge, methods and tools that must be designed and developed to produce and deliver services—may also advance sustainability practice. Thus, while environmental services address mainly "end of the pipe" issues, environmentally-friendly services and eco-efficient services aim to reduce the overall environmental impact of a solution by considering the upstream factors that may influence that impact. These can include the more efficient use of natural resources or the implementation of measures to cut discharge to the environment. Sustainable service, on the other hand, allows the customer to imbue the super-value of service with sustainability, which promotes a future-oriented approach that entails acting instead of reacting and that extends the profit boundaries beyond those of provider and customer while considering the long-term consequences at both the local and global scales, i.e., supplying sustainability by itself. Finally, equating sustainability with service allows the existing models, methods and tools developed in service science to be integrated in sustainability practice and sustainable development.

The framing of sustainability as a service is also in line with the core service characteristics mentioned in Chap. 2 [6, 7]. Thus, as in every service, sustainability is an *intangible* product, and sustainability as a value is produced and delivered simultaneously and jointly by the provider and the customer, i.e., *inseparability*. In addition, as sustainability cannot be stored and it is irreversible, it is also delivered in a definite time and place, and as such, it is *perishable*. Finally, as the concept of sustainability is complex, with high variability, it can never be repeated in exactly the same form, as the provider, the customer, and the time and the place vary from one delivery to the next, i.e., *heterogeneity* or inconsistency of a service. Yet despite its intangibility—sustainability conceptually encompasses and promotes moral values, such as justice and equity, it encourages awareness, and it propagates information, knowledge, and frameworks, methodologies and tools—the ultimate goal behind delivering sustainability as a service is to reduce the dependence on tangible natural, social and economic resources such as materials, energy, manpower, labor, effort, capital and entrepreneurship.

In contrast to the traditional services that we typically encounter on a daily basis, sustainability as a service must be produced and delivered in multiple dimensions. As such, the definitions of provider and customer and of the co-creation process are much broader and complex, because some of the actors are indirect and inactive providers or customers, and some of them do not even exist yet, i.e., future generations. In addition, the co-creation process herein integrates all service levels, from co-design through consume to co-produce, but at the same time, sustainability as a value should also produce and deliver multiple values in the environmental, social and economic dimensions. Producing and delivering sustainability as a service, therefore, also extends the above-mentioned characteristics of service—inseparability, perishability and heterogeneity—beyond the boundaries of the typical service, that is, from the individual to society and from local and short-term effects to include global and long-term outcomes. Additionally, sustainability as a service should provide a smart, continuous, dynamic and evolutionary value adapted to a certain place, time and people, e.g., habits, culture, class, and profession. Finally, note that there is no single service that alone can deliver sustainability as a whole, but a service that offers all the above-mentioned features together and that advances global sustainability can be considered a service that delivers sustainability as a value.

To summarize, a service that supplies sustainability as a value is actually a service that supplies the information, knowledge, framework, methods or tools as its core-value, which, in turn, increases awareness, action and responsibility toward sustainability as a super-value and eventually reduces the utilization of natural resources and damage to the natural and social environments. In addition, as sustainability is a complicated and amorphous concept, it is impossible to deliver it in a single service, but there are a variety of combinations of service types that can deliver all the requirements and aspects of sustainability. Such groupings of services can be found in many fields, from research on sustainability to fundraising for sustainability projects, from the design of frameworks for innovation to entrepreneurship in the field of sustainability, and from the development of methods or tools for sustainability assessments to the dissemination of information and knowledge about the practice of sustainability.

6.2 Design and Development of Sustainability as a Service

Due to the complexity and diversity of sustainability, the design and development of sustainability as a service should be based on multiple levels and dimensions. First, the direct and indirect needs that the service is intended to address must be defined and translated into direct and indirect basic values. Each of those values should then be assigned to provider/s and/or customer/s as stakeholders who assume responsibility for maintaining and propagating the values. Finally, the different basic values and their direct providers and customers should be organized into a value-chain that provides insight into the interactions between the basic values and into how and in which direction the entire supply-system should proceed.

6.2.1 Provider and Customer

As previously noted, sustainability as a service usually depends on many more than a single provider or customer. In addition, a service that delivers sustainability as a value is that in which each person can be a provider or a customer and, more importantly, every customer becomes a provider of sustainability. It means that sustainability as a value should consider a system comprising multiple providers that, in general, do not necessarily know each other, but their decisions and actions can affect the natural and social environment and thus humanity. Moreover, as the essence of sustainability is to live our lives today while providing enough space for the next generations to maintain their own lives, those future generations are effectively built-in customers of such a service. Finally, a customer that may benefit from the service may also be simultaneously a provider of the same value and eventually of sustainability. Practically speaking, the provider and the customer of sustainability as a service can vary from the individual level up to societies, from small to large social groups and from companies and firms to unions. This broad and comprehensive definition of providers and customers and their variety and large numbers represents one of the biggest challenges in the design of the co-creation process.

At last, although in the case of sustainability, theoretically, both provider and customer can be non-human entities, for example, nature or the economy, our primary methodology for the design and development of sustainability as a service focuses only on people as the providers and customers of services.

6.2.2 Value

As sustainability should integrate environmental, social and economic features and operate in several dimensions—e.g., from individuals/small groups/companies to societies/big groups/corporations, from short-term to long-term and from local to global levels—sustainability as a value should eventually incorporate all of these elements into a single, clearly defined value. Thus, to formulate the value-chain while considering the boundaries and dimensions of sustainability as a service we propose the following architecture (Fig. 6.1):

1. As each service comprises provider, value and customer, the first step in designing sustainability as a service is to break the **full-value** down into well-defined **basic-values** and to identify their direct and indirect providers and customers, i.e., defining **basic-service**.

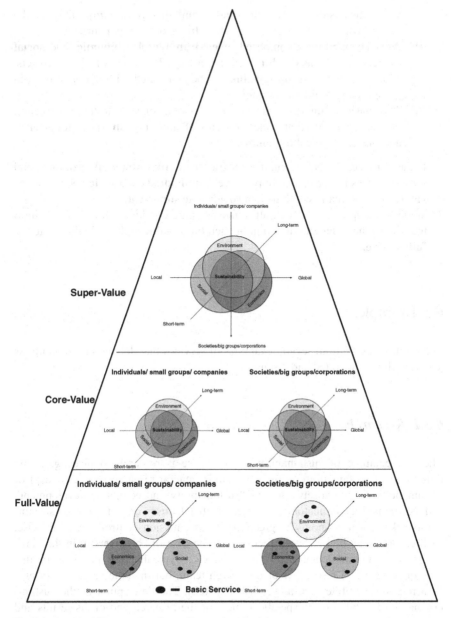

Fig. 6.1 Sustainability as value dimensions and methodology

2. The second step in designing sustainability as a service is to display each
   basic-value in its corresponding dimensions. This can be done as follows:

   (a) As the definitions of provider and customer are fundamental to every
       service, sustainability as a value is first divided into two levels, the first in

which the customers are **individuals, small groups** or **companies,** and the second where they comprise **societies, big groups** or **corporations**.

(b) As sustainability is composed of **environmental, economic,** and **social** aspects, each **basic-value** will be assigned to one of these aspects. Moreover, the same **basic-value** can be associated with one, two or even three aspects at the same time.

(c) The **basic-value** of each aspect is also displayed at the intersection between two different scales in terms of time, i.e., **short-** or **long-term,** and place, i.e., **local** or **global**.

3. In the third step of the design, the integration of **environmental, economic** and **social** services is aggregated in two levels, **individuals** and **societies,** i.e., **core-value,** that eventually combine to also generate **super-value**.

4. In the last step of the design, all dimensions are combined into a **value-chain** that shows the order and direction of each **basic-value** and that builds up to the **full-service**.

6.3 Examples

In the following section, several examples of services that deliver sustainability as a value will be illustrated and discussed.

6.3.1 Research

The first example of sustainability as a service is research to investigate any dimension of sustainability. This can include general research of the concept of sustainability in its entirety, such as "sustainability and economic development" and "sustainability indicators", or research that focuses on a feature of sustainability, like "local economy", "greenhouse gas emission control", etc. Likewise, such research can be narrow in focus, as in the case of "biodiversity in the black forest" or "environmental activism in Israel". Typically, the provider in this type of service is a researcher or a research team, but in some cases, it can also be a research institute or a university (Fig. 6.2). The description of the customers mainly depends on the specific subject of the research and on its results and insights. However, as the concept of "customer" is highly diverse, its actual definition can vary widely, from other researchers and experts to decision-makers or even laypeople (Fig. 6.2).

The direct value of every research endeavor can be generally defined as knowledge. This knowledge can usually be translated—either locally or globally and for the short- or long-term—into awareness and decision-making processes as part of the service core-value. However, it can also promote the production and delivery

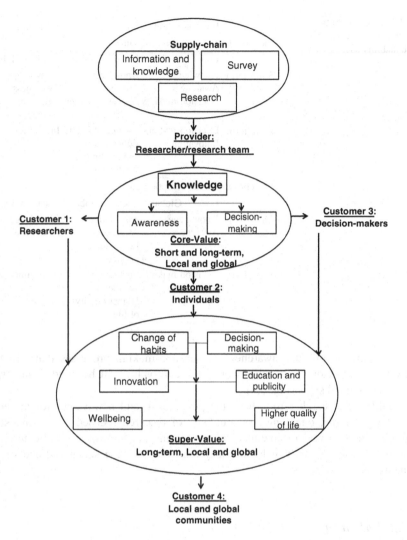

Fig. 6.2 Value-chain of research as an example of sustainability as a service

of other actions that will affect individuals and societies in the long term, from changes in habits and in decision-making processes through innovation, education and publicity to wellbeing and providing a higher quality of life (Table 6.1 and Fig. 6.2).

Figure 6.2 illustrates the organization of the main basic-values into the value-chain using the example of research. Of course, there are also other basic-values that can be added, and other customers or providers may also take part in the value co-creation process via other pathways. For example, the core-value can be provided in a different fashion whereas the direct-value that the service provides, i.e., knowledge, will be used by experts, governmental or non-governmental

Table 6.1 Basic-values of research as an example of sustainability as a service

	Environmental		Social		Economic	
Individuals/small groups/companies						
	Local	Global	Local	Global	Local	Global
Short-term	Knowledge Awareness Decision-making		Knowledge Awareness Decision-making		Knowledge Awareness Decision-making	
Long-term	Decision-making Innovation		Change of habits Education and publicity Wellbeing	Innovation		
Societies/big groups/corporations						
	Local	Global	Local	Global	Local	Global
Short-term	Knowledge Awareness		Knowledge Awareness Education and publicity		Knowledge Awareness	
Long-term	Decision-making Innovation		Innovation Wellbeing Higher quality of life		Innovation	

organizations, to produce awareness or decision-making processes that will be delivered to individuals. In other words, the core-value will be divided into two additional defined steps (Fig. 6.3).

Finally, all research efforts have the potential to add new values for the end user. For example, one long-term outcome of research focused on biodiversity could be the discovery of new medications for and medical solutions to emerging viral or disease threats, which can yield values in both the social and economic dimensions.

6.3.2 Labeling

As previously discussed in Chap. 5, the ranking or labeling of goods or companies with regard to environmental performance is a growing field. A pioneering service of this type was introduced in 2006 by The Carbon Trust, a UK company [8] that performs carbon footprint assessments and labeling. A goods carbon labeling service, it enables customers to compare different goods or brands from an environmental perspective.

In general, a labeling service has three direct customers: the producer of a good and its marketers and end consumers. While the direct-value of the service is the labeling itself, in fact, the service is a tool for providing knowledge and awareness, with the ultimate goal of effecting action that includes, but is not limited to, changes in behavior. The core-value of the service, therefore, should also include

Fig. 6.3 Alternative value-chain of research as an example of sustainability as a service

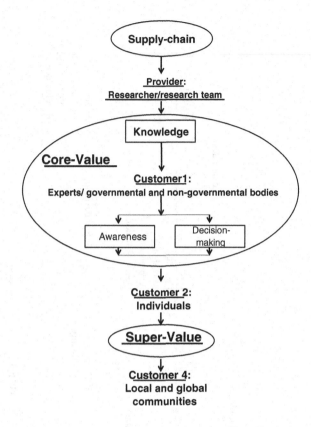

a more environmentally friendly decision-making process that considers the production and purchase of a good, which should be guided by the overall goal of reducing carbon emissions (Table 6.2 and Fig. 6.4). However, in the long term and from a much broader perspective, the use of carbon labeling—besides its potential indirect beneficial effects, such as decreased carbon emissions and reductions in resource utilization—can and should ultimately encourage its customers to change their consumer habits. For example, a labeling service can initiate fundamental changes in the culture of consumption typical of the West (but today, more of a global phenomenon) and promote innovation and more healthy competitiveness. This will create local economies through a co-creation process of shared economic development that, in turn, leads to increased stakeholder engagement and, eventually, increased wellbeing, prosperity and a higher overall quality of life.

6.3.3 Online Monitoring and Presentation System

The final example of sustainability as a service discussed here is an online monitoring and presentation system that produces and delivers real-time information through a value co-design process. A transit information system such as Tiramisu

Table 6.2 Basic-values of carbon labeling as an example of sustainability as a service

	Environmental		Social		Economics	
	Local	Global	Local	Global	Local	Global
Individuals/small groups/companies						
Short-term	Labeling Knowledge Awareness Decision-making Decreased carbon emissions		Labeling Knowledge Awareness Decision-making		Labeling Knowledge Awareness Decision-making	
Long-term	Decreased carbon emissions Reduction in resource use		Change of habits Wellbeing	Wellbeing Higher quality of life	Local economy development	Higher competitiveness Share economy development
Societies/big groups/corporations						
Short-term			Innovation Change of consuming culture		Efficiency-profit	
Long-term	Innovation Decreased carbon emissions Reduction in resource use		Innovation Change of consuming culture	Change of consuming culture	Innovation Local economy development	Share economy development Prosperity

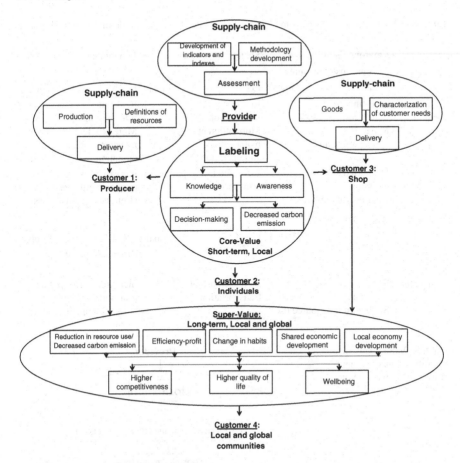

Fig. 6.4 Value-chain of carbon labeling as an example of sustainability as a service

(discussed in Sect. 5.5) that provides real-time bus tracking information is an example of such service. Its sustainability features are described and illustrated in Table 6.3 and Fig. 6.5 [9].

Because the primary service provider (the bus company) is in charge of the service delivery platform, the provider of a real-time bus tracking service can be the bus company itself, or the tracking service alone could be outsourced by the bus company or run by an independent company, e.g., a mobile application offered by a private company. The customers in this example are the bus passengers, who can plan their schedules using the published bus routes, and the bus company and its drivers, who can manage the schedules and routes differently. But because not only the bus drivers and the bus company, but also the passengers, are feeding information into an online system, all three actors are, in fact, providers of the same value to other customers.

The service core-value in this case, of course, is real-time information, but the implementation of a bus tracking service has the inherent potential to create more

Table 6.3 Basic-values of a real-time bus tracking system as an example of sustainability as a service

	Environmental		Social		Economics	
Individuals/small groups/companies						
	Local	Global	Local	Global	Local	Global
Short-term	Real-time information More efficient management and operation		Real-time information More efficient management and operation		Real-time information More efficient management and operation	
Long-term	Decreased carbon emissions Cut of resources use		Increased use of public transportation		Increased public transportation mileage Higher competitiveness	
Societies/big groups/corporations						
	Local	Global	Local	Global	Local	Global
Short-term					Efficiency-profit	
Long-term	Decreased carbon emissions Reduction in resource use		Increased use of public transportation		Increased use of public transportation Higher competitiveness	

Fig. 6.5 Value-chain of a real-time bus tracking system as an example of sustainability as a service

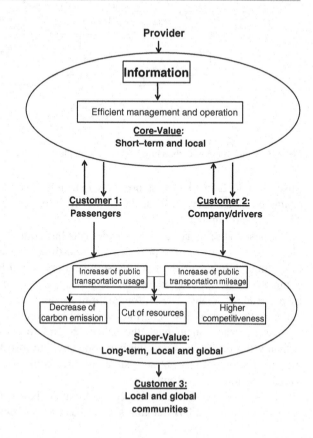

efficient management and operation of the bus system. In addition, in the long term the success of the system can lead to a greater acceptance and use of public transportation, subsequently increased public transportation mileage (with a corresponding reduction in private car use), a more competitive bus company, and, eventually, decreased carbon emissions and reductions in resources used by the bus company.

References

1. Willard B (2002) The sustainability advantage. New Society Publishers, Gabriola Island
2. Dresner S (2008) The principles of sustainability, 2nd edn. EarthScan, Oxford
3. Lusch RF, Vargo SL (2006) The service-dominant logic of marketing: dialog, debate, and directions. M.E. Sharpe Inc., New York
4. Vargo SL, Lusch RF (2008) Service-dominant logic: continuing the evolution. J Acad Mark Sci 36(1):1–10
5. Vargo SL, Maglio PP, Akaka MA (2008) On value and value co-creation: a service systems and service logic perspective. Eur Manage J 26:145–152
6. Regan WJ (1963) The service revolution. J Mark 47:57–62
7. Rathmell JM (1966) What is meant by services? J Mark 30:32–36
8. http://www.carbontrust.com. Accessed 25 May 2014
9. http://www.tiramisutransit.com. Accessed 25 May 2014